Organic Chemistry

Human benefit from the organic chemistry of plants is incalculable in terms of health, food, comfort and security. Indeed, the future well-being of humanity rests in significant measure upon a responsible relationship with the plant kingdom in order to re-establish balance in the Earth's natural environmental systems.

In a highly readable volume, *Organic Chemistry: Miracles from Plants* presents many fascinating points of entry to the organic chemistry of a wide range of crucially-important, naturally-occurring, chemical substances which are derived from plants.

Features:

1. Presents in a readable and accessible manner many fascinating points of entry to the organic chemistry of a wide range of crucially-important, naturally-occurring, chemical substances which are derived from plants.
2. Key concepts in and knowledge of organic chemistry are reinforced.
3. Highly-relevant and contemporary context stimulates learning in organic chemistry.
4. Searching exercises and extension materials are provided at the end of every chapter each of which is amply illustrated.
5. In a single source, this volume provides knowledge, challenge and valuable learning opportunity in chemistry, medicine, nutrition and the environmental sciences.

Organic Chemistry

Miracles from Plants

By
Jeffrey J. Deakin

CRC Press
Taylor & Francis Group
Boca Raton London New York

CRC Press is an imprint of the
Taylor & Francis Group, an **informa** business

First edition published 2024

by CRC Press
2385 Executive Center Drive, Suite 320, Boca Raton, FL 33431

and by CRC Press
4 Park Square, Milton Park, Abingdon, Oxon, OX14 4RN

© Jeffrey J. Deakin

CRC Press is an imprint of Taylor & Francis Group, LLC

Library of Congress Cataloging-in-Publication Data
Names: Deakin, Jeffrey John, author.
Title: Organic chemistry : miracles from plants / by Jeffrey J. Deakin, The Royal Society of
   Chemistry, London.
Description: Boca Raton ; London : CRC Press, 2024. | Includes bibliographical references
   and index.
Identifiers: LCCN 2023055041 (print) | LCCN 2023055042 (ebook) | ISBN 9781032664910 (hbk) |
   ISBN 9781032664903 (pbk) | ISBN 9781032664927 (ebk)
Subjects: LCSH: Chemistry, Organic.
Classification: LCC QD251.3 .D43 2024 (print) | LCC QD251.3 (ebook) | DDC 547—dc23/
   eng/20240129
LC record available at https://lccn.loc.gov/2023055041
LC ebook record available at https://lccn.loc.gov/2023055042

ISBN: 978-1-032-66491-0 (hbk)
ISBN: 978-1-032-66490-3 (pbk)
ISBN: 978-1-032-66492-7 (ebk)

DOI: 10.1201/9781032664927

Typeset in Times LT Std
by Apex CoVantage, LLC

Contents

PART I   Introduction

PART II   Medical Marvels

PART III Nutrition—Ancient and Modern Miracles

PART IV   Beverages

PART V   Euphorics

PART VI   *Exotic Potions, Lotions and Oils*

PART VII   Colorful Chemistry, a Natural Palette of Plant Dyes and Pigments

PART VIII   Plant Materials

PART IX   Plants and the Natural Environment

Preface

This book concerns the study of organic chemistry in the context of diverse substances from plants that are of huge significance in the contemporary world.

Organic Chemistry: Miracles from Plants provides ample extension material to enrich knowledge, deepen understanding, strengthen appreciation of organic chemistry and provide inspiration for further study. The chemical compounds from plants that are described are associated with important drugs, foods, beverages, perfumes, cosmetics, pigments, materials and contemporary issues facing society.

This book will be of particular value to students who are in the upper age groups of high school or at early university level. Reinforcement of the relevance and importance of organic chemistry in the modern world offers powerful stimulus to motivate students and will enhance learning.

When used in the broader sphere of general studies classes, extracts selected by tutors from a wealth of cross-curriculum material may stimulate informed debate about the relationship between scientific development and commercial exploitation of the products of plant biochemistry fostering exploration of related business, ethical and social issues.

Furthermore, the material is also intended to excite the interest of scientifically literate people who wish to broaden their horizons on the basis of personal or professional motivation.

Author Biography

Dr Jeffrey J. Deakin earned a BSc degree in chemistry with first class honors from the University of London followed by a PhD degree in physical chemistry from the University of Cambridge.

As a teacher of science, Jeff headed chemistry and physics departments in grammar and comprehensive schools in the UK. He has written books and numerous articles aimed at demystifying chemistry and broadening interest in the subject.

Jeff is a Fellow of The Royal Society of Chemistry in London. He was a member of the Curriculum and Assessment Working Group at the Royal Society of Chemistry, which recently reviewed the national curriculum in chemistry in each of the four home nations of the United Kingdom of Great Britain and Northern Ireland.

Part I

Introduction

ORGANIC CHEMISTRY

In 1807, the Swedish chemist, Berzelius, considered organic compounds to be chemical substances derived from living organisms and inorganic compounds to be those obtained from inanimate matter. It was believed that organic compounds could only be produced through a' vital force' inherent in living cells and therefore could not be synthesized artificially.

However, in 1828, Woehler, a German chemist, showed that heating a compound hitherto regarded as inorganic, ammonium cyanate, produced a compound hitherto regarded as organic, urea:

$$NH_4NCO = CO(NH_2)_2$$

The 'vital force' theory was undermined and was finally disproved by Kolbe shortly thereafter who made acetic acid (ethanoic acid) directly from its constituent elements: carbon, hydrogen and oxygen.

Organic compounds are now defined as carbon compounds without reference to their source or to how they were produced.

Organic chemistry is essentially the study of the chemistry of life—illustrated in this book by products of and extracts from plants that are of significance to man.

Although metallic elements considerably outnumber non-metallic elements in the Periodic Table (Figure 1.1), the versatility of carbon ensures, through its ability to bond repeatedly to itself, that organic chemistry dominates the field of chemistry.

Organic chemistry is extensive because a carbon atom, having a valency of four, may be bound covalently to another carbon atom through a single bond, a double bond or a triple bond. Significantly, carbon atoms can bond together to form different structures, such as chains and rings. Carbon can also combine covalently with other elements, typically hydrogen, oxygen, nitrogen, sulfur and phosphorus. As a result, infinite combinations and permutations are possible, which lead to an apparently bewildering array of naturally produced organic molecules: some simple and others complex although all of them are fascinating.

It is important reassure the reader at this point that he/she does not need to acquire detailed knowledge of every single organic compound for they can be divided into various classes, each containing a characteristic group of atoms that have particular chemical properties.

DOI: 10.1201/9781032664927-1

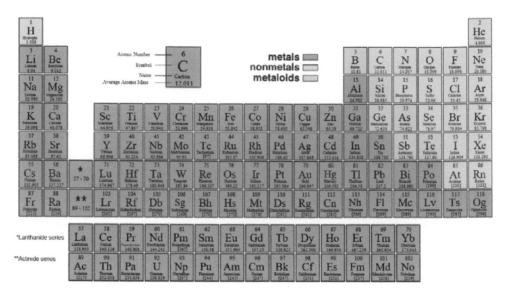

FIGURE 1.1   Periodic table of elements, metals and non-metals.

Source: https://commons.wikimedia.org/wiki/File:Periodic_Table_Of_Elements.svg

The naturally produced chemical substances illustrated are described in relation to the basis of the organic chemistry curricula for high school and early university courses in the United States of America and for sixth form and early university courses in the United Kingdom. The aims are to inspire, to inform and to extend the understanding of students while helping them to gain an appreciation of the organic chemistry involved.

Although the scope of the book is comprehensive, it has not been the intention to cover the curriculum systematically as an instructional textbook might do. Nonetheless, study of an extensive range of concepts and functional groups is interwoven in the text:

- covalent bonding
- alkanes, linear and branched
- halo alkanes
- cycloalkanes
- compounds containing the carbonyl group
- alcohols
- esters
- nitriles
- alkenes
- alkynes
- aromatic chemistry (arenes)
- amines and amides
- amino acids (including the peptide link and proteins)
- phenols
- heterocyclic compounds
- polymers
- isomerism
- optical isomerism
- chirality.

The chemistry of functional groups is drawn upon extensively to show how they influence the chemical behavior of the building blocks that make up large and complex molecules. A building block is a term used in organic chemistry to describe a part of the structure of a large molecule that has one or more active functional groups. Examples of building blocks from later in the book are isoprene, phenol, amino acids and sugars.

Different building blocks may be assembled, through reactions involving the functional groups, to form much bigger or more complex molecules. Organic chemists then divide compounds with large molecular structures into categories that have a common chemical background. Examples of these categories are carbohydrates, terpenoids, steroids, proteins, polyphenols, saponins, lipids (fats and waxes), carotenoids, sugars (polysaccharides) and fibers. These categories are examined.

The roles of isomers and polymers are fully explained since stereo-chemistry subtly and profoundly influences the interactions of large, complex molecules found in nature and hence their physical and chemical properties.

There is an introduction to techniques that are used to investigate the structure of large molecules and to the application of chromatography in purification and analysis.

For the instructor, the text supplements the core curriculum and offers inspiration and materials to inform and support project work by students working individually or in groups. Teachers and tutors may lead by choosing materials that will extend and illuminate the curriculum tailored both to the specific requirements of the syllabus and to the level and interest of their students.

CONTEXT: MIRACULOUS CHEMICALS FROM PLANTS

Plants have changed the world influencing civilizations, trade and conquest.

Important examples of the discovery of valuable medicinal plants are presented in **Part II** that were of inestimable benefit to mankind. It might be said that many people in the world only know of indigenous medicines based upon ancient tradition, yet we in the West are recipients of these wonderful gifts without perhaps ever appreciating their origins.

Chemical analysis of plant extracts has led to the identification of many natural products of considerable value in modern foods and beverages. These are described in **Part III** and in **Part IV**. A powerful example was the use of lime fruits by the British navy, which overcame scurvy even though it took another 200 years to discover the reason was the presence of vitamin C in the citrus plant.

In the 19th century, chemical compounds were identified as poisons or capable of affecting mood. Some of these are described in detail in **Parts V** and **VI**. Two powerful chemicals, morphine and strychnine, were isolated from different plants in 1815 and 1819, respectively, although their actual chemical structures were not defined for another 100 years! Numerous examples of pharmaceutical and over-the-counter drugs were derived directly from compounds found in nature or indirectly through chemical modification of the basic chemical structure.

We appreciate the beauty of color in the natural environment which inevitably influences art and science through the exploitation of natural pigments. Therefore, **Part VII** highlights the colorful chemistry of natural products. Have you ever wondered what makes the world of plants so colorful? Why are the leaves green and then turn to wonderful hues of yellow and red in the autumn? The explanation is due to the presence of natural pigments in the plants. Indeed, historically, many pigments from plants were used as dyes and these were revered also for possessing 'magical properties' with the power to heal and to keep evil spirits at bay. Mystery and superstition surrounded the extraction of the essence. An example of this is provided in the use of *Woad*. This is a plant, *Isatis tinctoria*, which yields the purple-blue color, indigo, scarce in the natural world. The desire and need for different colored dyes stimulated much research by chemists in the 19th century that contributed to transformation of society from its historic agrarian foundation to the modern industrial era.

In **Part VIII**, the vital importance of all the plants of the natural world is reinforced. Directly and indirectly, humankind and other members of the Animal Kingdom are wholly dependent upon plants. The Plant Kingdom is dominated by green plants converting carbon dioxide and water into a vast and diverse array of organic compounds, not only providing vital resources in food but also for clothing and shelter and releasing indispensable oxygen into the atmosphere for respiration.

Attention is drawn to plants as providers of a wide variety of materials, instances being paper products, fabrics used in clothing and those of high strength offering tools and shelter.

In the concluding **Part IX**, we consider the inestimable value and highly significant influence of plants upon the well-being of the natural chemical cycles and inter-dependent systems of planet Earth.

REFERENCES

Selected reading material is included for those wishing to delve further into the background. Primary references are cited which support key points.

NOMENCLATURE

Many natural chemical products are given trivial names and these are used throughout. However, where appropriate, the IUPAC system is adopted to describe chemical formulae.

BOTANICAL NAMES

Every plant belongs to a botanical family, which is further classified into its genus and its species. The taxonomy of the plant is given a Latin name expressing both the genus and species and is italicized.

SCOPE OF CHEMISTRY

For ease of reference, the scope of the organic chemistry presented in this book has been tabulated at the beginning of each section and chapter.

Part II

Medical Marvels

INTRODUCTION

Medicines derived from plants are widely used in traditional cultures all over the world. People who use traditional remedies may not understand the scientific rationale for why they work but know from personal experience that some plants can be highly effective.

Traditional medicine often aims to restore balance to the body. Western allopathic medicine, in contrast, often uses a well-defined, single chemical entity with specific medicinal properties. Nevertheless, many of our so-called modern day drugs have origins in ancient medicine.

The vast array of medicinal plants available from all parts of the world has stimulated much scientific and clinical interest which, in some instances, has provided significant commercial returns and modern medications of inestimable value to humankind.

A summary of the chemistry in Part II follows.

Chapter	Organic Chemistry	Context
Humble potato	Alkanes, cycloalkanes	Steroids
	Benzene and aromaticity	Hormones
	Industrial fractional distillation	
Willow Bark	Carboxylic acids	Salicylic acid and aspirin
	Phenol	
	Acetylation	
	Substituted aromatic compounds	
Cinchona and Artemisia	Carbon–oxygen bonds	Quinine
	Oxygen–oxygen bonds	Artemisinin
	Peroxides	
Foxglove	Alcohols, Acids, Esters	Glycoside lactones (cyclic esters)
Periwinkle	Functional groups	Alkaloids
	Fractional distillation	Indole alkaloids
	Acid—Base extraction	Vincristine and vinblastine

(Continued)

DOI: 10.1201/9781032664927-2

(Continued)

Chapter	Organic Chemistry	Context
Pacific Yew	Isomers	Terpenes
	Elimination reactions	
	Stereo-chemistry	
	Chirality	
	NMR spectroscopy	
Vaccines	Alkenes	Saponins
Adjuvants	Carbohydrates	Chilean Soapbark Tree
	Proteins	RNA and DNA
	Glycosides	
	Isomers	

CENTRAL AMERICA'S HUMBLE POTATO!

Abstract: How a humble potato led to the genesis of the birth control pill. Few chemical extracts have had a greater impact on modern society. The potato is indigenous to Central America and was originally used as a staple food. The tuber of the potato contains chemical compounds that led to the transformation of the world through the development of the modern birth control pill, bringing about the profound social, cultural and economic impacts of oral contraception.

Organic Chemistry

- *hydrocarbons*
- *alkanes and cycloalkanes*
- *benzene and aromatic compounds*

Context

- *steroids*
- *hormones, estrogen and progestogen.*
- *the profound social, cultural and economic impacts of oral contraception.*

STEROIDS

A steroid contains a characteristic arrangement of four cycloalkane rings that are joined to each other. The main way in which steroids vary from one another is through the functional groups attached to the four-ring core (**Figure 2.1**).

Hundreds of distinct steroids are found in plants, animals and fungi. Examples of steroids include the dietary fat, cholesterol and the sex hormones, testosterone and progesterone (**Figure 2.2**).

THE SEARCH FOR THE MODERN BIRTH CONTROL PILL

The modern, oral contraceptive pill, often referred to as the birth-control pill or colloquially as 'the pill', is a birth control method that involves a combination of hormones, an estrogen and a progestogen. When taken orally every day, these pills inhibit female fertility. They were first approved for contraceptive use in the United States in 1960 and have become a very popular form of birth control used by many millions of women worldwide.

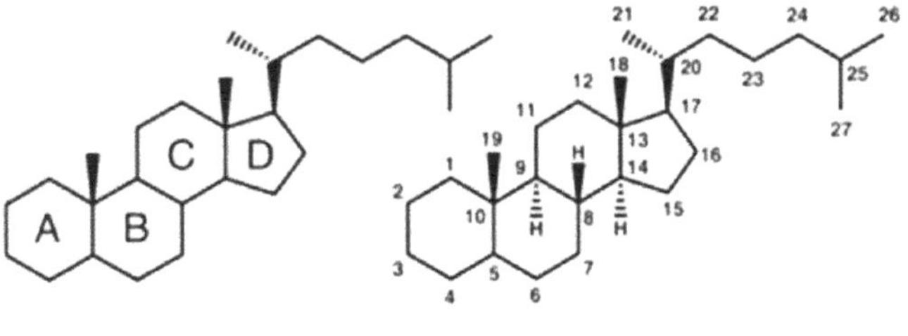

FIGURE 2.1   The carbon skeleton and ring structure of steroids together with the numbering system for steroids.

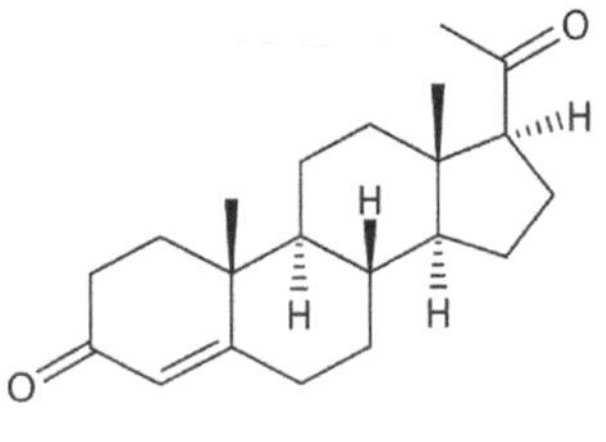

FIGURE 2.2   Chemical structure of progesterone.

The Mexican potato (*Dioscarea villosa* or *Dioscarea barbasco*) was a key and early commercial source of important chemical compounds known as steroid saponins. The wild potato itself has no contraceptive value. Essentially, it provides the starting material from which the steroidal hormones are manufactured by modern technology through a combination of synthetic methods and also microbial transformation. However, in the 1930s, before the discovery of the usefulness of this potato, two important discoveries were made.

Firstly, scientists isolated and determined the structure of small amounts of steroid hormones. They then found that high doses of these hormones, in the form estrogens or progesterone, inhibited mammalian ovulation. The isolation of these natural hormones was achieved by research on animal sources and carried out by the major European pharmaceutical companies. Since only small amounts were recovered they were extraordinarily expensive. However, it became immediately apparent that an urgent need for an abundant, reliable and cheap supply of steroids was needed.

ISOLATION OF NATURAL DIOSGENIN FROM THE MEXICAN POTATO

In 1939, Professor Russell Marker, at Pennsylvania State University, developed a method of synthesizing progesterone from plant steroids, which are known as sapogenins.

This led to the investigation of a variety of Mexican potato, including *Dioscorea mexicana* and *barbasco*, which are found in the rain forests of Veracruz, for the tuber (root) of the potato contains the natural chemical called diosgenin (**Figure. 2.3**). This abundant compound could be used as the starting material in the synthesis of hormones on an industrial scale. Diosgenin can be easily converted chemically by opening the rings containing C21 to C27 as shown in **Figure. 2.1**. This is then followed by further degradation of the molecule to reach the structure shown in **Figure. 2.2**.

Although by midway through the 20th century the stage appeared set for the development of an oral hormonal contraceptive, pharmaceutical companies, universities and governments showed little interest in pursuing further research. At this time in 1944, having developed a synthesis of progesterone from diosgenin from the Mexican potato, Professor Marker left Pennsylvania State University to found a new company, Syntex, in Mexico City. At Syntex, Marker continued to perfect the extraction of the saponins from *Dioscorea mexicana* and then the manufacture of the key hormones. Importantly, due to this achievement, the monopoly of the European pharmaceutical companies was broken, which had until that time controlled production of steroid hormones. As a consequence, the price of progesterone fell dramatically by almost 200-fold over the next 8 years.

CHEMICAL MAGIC IN THE LABORATORY—SYNTHESIS OF NORETHINDRONE

In 1951, the combined brilliance of three extraordinary chemists at the Syntex company in the USA (Carl Djerassi, Luis Miramontes and George Rosenkranz) led to the synthesis of the first oral progestin, namely, norethindrone (**Figure 2.4**). This synthetic hormone is a variation of natural progesterone.

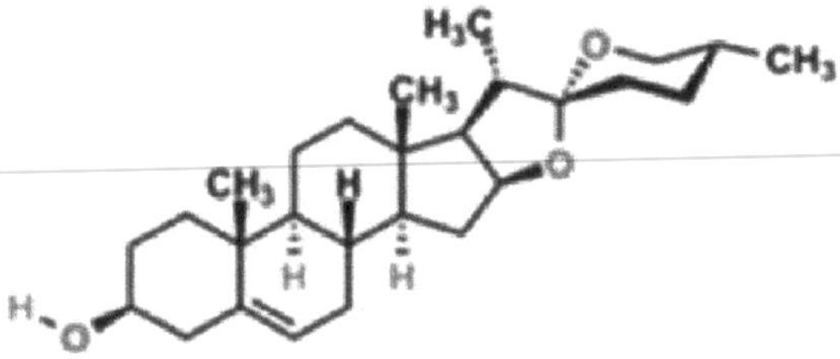

FIGURE 2.3   The chemical structure of diosgenin.

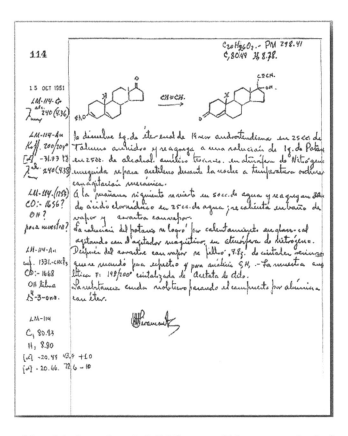

FIGURE 2.4 A copy of the original page from Luis E. Miramontes' laboratory notebook, signed October 15, 1951.

BIOLOGICAL STUDIES OF PROGESTERONE TO PREVENT OVULATION

Stimulated by a steady supply of research grade chemical compounds, important biological studies on the activity of progesterone in inhibiting ovulation progressed rapidly. In early 1951, the reproductive physiologist Dr. Gregory Pincus, who was at the time a leader in hormone research, met with Margaret Sanger, the founder of the American birth control movement (see historical note).

HISTORICAL NOTE: THE IMPACT OF MARGARET SANGER

The importance of Margaret Sanger, a nurse and a huge advocate of female contraception, cannot be overlooked. Her crusade to legalize birth control spurred the movement for women's liberation.

Margaret Sanger became acutely aware of the effects of unplanned and unwelcome pregnancy during her work with poor women on the Lower East Side in New York. Sanger had witnessed how her own mother's health had suffered as she bore eleven children. Sanger appreciated the importance to women's lives and health of the availability of birth control. In 1912, Sanger gave up her nursing work to dedicate herself full time to the distribution of birth control information. During World War I, Sanger set up the first birth control clinic in the United States, yet she was arrested and prosecuted many times. The resulting public outcry helped lead to changes in the law, which in turn empowered doctors to give birth control advice to their patients.

In 1927, Sanger helped to organize the first World Population Conference in Geneva.

In 1942, after several organizational mergers and name changes, The Planned Parenthood Federation (PPFA) came into being.

MEDICAL APPROVAL AND SOCIAL ACCEPTANCE

By 1954, studies had advanced on the ovulation-suppressant potential of progestins that were administered orally. The first medical trials of an oral contraceptive, later commercially known as Enovid, began in 1956 in Puerto Rico, which led the American Food and Drug Administration (FDA) to approve Enovid for menstrual disorders, and by May 1960, the FDA gave approval for its use as a contraceptive.

THE PROFOUND SOCIAL, CULTURAL AND ECONOMIC IMPACTS OF RELIABLE ORAL CONTRACEPTION

Since the introduction of the oral birth control pill in 1960 and its approval by the FDA, its use has spread rapidly generating enormous social impact—a significant factor in a quiet revolution. Many consider it to be the most socially significant medical advance of the 20th century. The birth control pill has helped women gain more control of their lives and has altered the nature of the nuclear family and life profoundly. Many economists argue that the availability of the birth control pill led directly to an increase in the proportion of women in the labor force and was a key influence in determining the modern economic role of women. Family planning allowed women to make long-term educational and career plans. The pill offered opportunity to delay the timing of marriage, allowing women to invest in education and in other forms of human capital and to become more career-oriented.

Due to the fact that the birth control pill was inexpensive and effective, widespread adoption changed the nature of debate over pre-marital sex and promiscuity. Never before had sexual activity been so divorced from human reproduction. In this regard, the proliferation of the use of oral contraceptives has required religious authorities to address the ethical relationship between sexuality and procreation.

TRANSFORMATION OF PRODUCTION

In 1960, 2 million women were using the pill and over 100,000 Mexican peasants were gathering the raw material used in its production. In order to meet demand more than 10 tons of wild potatoes were removed each week at extraordinarily low prices from the areas around Oaxaca, Veracruz, Tabasco and Chiapas in Mexico. Scientists relied on local, indigenous knowledge to cultivate and harvest the plant. Potatoes made their way from the Mexican jungles to domestic and foreign laboratories and into the medicine cabinets of millions of women around the world. At the time, little recognition was afforded to Mexican peasants who labored for almost 30 years to collect the potatoes yet had no sense of its value in the marketplace or of the importance of their contribution.

Interest in the Mexican potato could no longer be confined within national borders as growing pressure arose from a combination of continued progress in chemistry research, improved pharmaceutical technology and changes in the social and political outlook across the world. The Mexican government eventually established a state-owned company in 1975 to compete with foreign laboratories. Funds were thus secured for the training of scientists and the development of a stronger domestic pharmaceutical industry in Mexico.

Arguably, the indigenous, poor, uneducated potato pickers represented in many respects the antithesis of modernity, but they became an essential link in finally introducing to Mexico modern, domestic industry-patented medications. In this particular case, an alliance of science and farming practice resulted in a reshuffling of social hierarchy in rural Mexico and gave real monetary value to an otherwise low value crop

HYDROCARBONS

Table 2.1 illustrates the general classification of hydrocarbons and provides examples.

TABLE 2.1

The General Classification of Hydrocarbons

Hydrocarbons						
Chain			Cyclic			
Saturated	Unsaturated		Carbocyclic		Heterocyclic	
Alkanes	Alkenes	Alkynes	Ali-cyclic	Aromatic	Ali-cyclic	Aromatic
(methane)	(ethene)	(ethyne)	(cyclohexane)	(benzene)	(cyclohexylamine)	(pyridine)

SATURATED HYDROCARBONS—ALKANES AND CYCLOALKANES

Alkanes

Alkanes are described as saturated hydrocarbon molecules because each carbon atom, which has a valency of four, is bound covalently to four other atoms—either carbon or hydrogen.

Alkane molecules can be in straight chains of carbon atoms (**see examples in Table 2.2**) or in branched chains.

TABLE 2.2

Examples of Hydrocarbons with the General Formula C_nH_{2n+2}, Known as Alkanes

Formula	Name	Physical Properties	Boiling Point	Melting Point
CH_4	methane	colorless gas		
C_2H_6	ethane	colorless gas		
C_3H_8	propane	colorless gas		
C_4H_{10}	n-butane	colorless gas		
C_5H_{12}	n-pentane	colorless, volatile liquid	36°C	
$C_{18}H_{38}$	n-octadecane	white solid		28°C

Cycloalkanes

Cycloalkanes exist too, although they are usually in the form of six-membered carbon rings or larger. Due to the ring structure, however, some strain exists in the bonds. For instance, flat planar molecules of cyclopropane can be formed but they are very unstable, whereas in cyclohexane (**Figure 2.5**) the carbon-carbon bonds are formed at the usual tetrahedral angle of 109.5 degrees. As a consequence, the cyclohexane ring is puckered into what is commonly referred to as either the chair or the boat form.

FIGURE 2.5   The structure of a cyclohexane molecule.

Owing to the ring structure, cycloalkanes have two fewer hydrogen atoms than the corresponding alkane and have the general formula C_nH_{2n}. However, the physical and chemical properties of cycloalkanes are scarcely different to those of the corresponding straight chain or branched chain alkane.

Chemical Properties

Alkanes and cycloalkanes, in gaseous or vapor form, are readily oxidized in the presence of oxygen or air with the release of a great deal of heat energy per mole.

$$C_3H_8 + 5O_2 = 3CO_2 + 4H_2O$$

A mole is defined as the molecular weight of a substance expressed in grams based on the standard of 12 grams of carbon 12.

In the presence of a halogen and energized by light, alkanes undergo substitution reactions

$$CH_4 + Br_2 = CH_3Br + HBr \text{ then } CH_3Br + Br_2 = CH_2Br_2 + HBr$$

and so on, to CBr_4.

CRUDE OIL AND INDUSTRIAL FRACTIONAL DISTILLATION

Crude oil or petroleum is a highly complex mixture of alkanes and many other organic compounds. Petroleum is found in underground reservoirs in rock strata where it was formed by the very slow decomposition of organic matter from plants and animals in the absence of air under the influence of great heat and pressure. The composition of crude oil varies considerably from place to place; some deposits being dominated by straight and branched chain alkanes whereas other sources contain greater proportions of cyclic alkanes that often have carbon chain branches.

Crude oil is fractionally distilled and different fractions are collected on an industrial scale (Table 2.3) in oil refineries (Figure 2.6).

BENZENE AND AROMATIC COMPOUNDS—PHYSICAL AND CHEMICAL PROPERTIES

Benzene is a colorless, volatile liquid and is a recognized carcinogen.

The molecular formula, C_6H_6, reveals that there is a high percentage of carbon and that benzene is to some degree unsaturated. Indeed, benzene is described as the simplest of the aromatic hydrocarbons. Aromatic hydrocarbons are also known as arenes. The reference to aromatic arose from the pleasant smells of the earliest discovered arenes. The simplest arene is benzene itself, C_6H_6. The next simplest is methylbenzene (old name: toluene) which has one of the hydrogen atoms attached to the ring replaced by a methyl group—$C_6H_5CH_3$. Then there are polyaromatic hydrocarbon compounds, such as naphthalene (C10H8) with a further carbon ring and so on in complexity.

TABLE 2.3
Examples of Oils That Can be Refined by Fractional Distillation

Fraction	Boiling range	Carbon atoms per molecule	Use
Light petroleum	20–90°C	4 to 6	solvent
Gasoline (petrol)	100–200°C	8 to 12	motor fuel
Paraffin, kerosene	200–300°C	12 to 16	diesel fuel
Oil	above 300°C	more than 25	lubrication
Bitumen	solid residue	large numbers	road construction

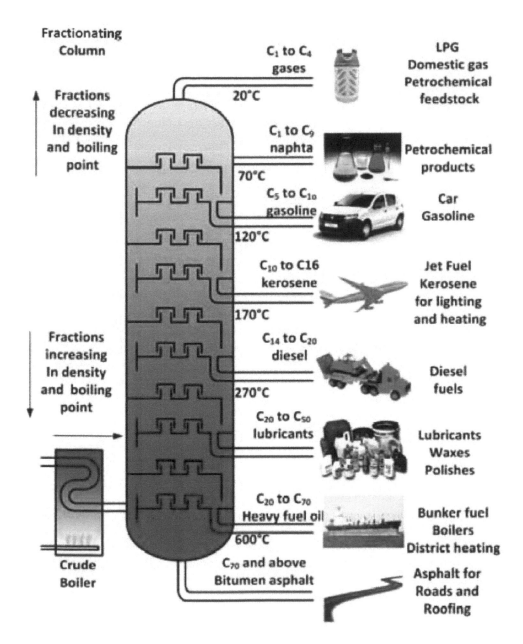

FIGURE 2.6   Diagram of an industrial fractionating column.

Studies of the structure of a molecule of benzene have shown that the molecule is in the form of a planar, hexagonal ring of six carbon atoms with the internal angle between them being 120 degrees. This is in marked contrast to the molecular structure of cyclohexane, which is discussed earlier in the chapter. Nowadays, the terms aromatic or arene also indicate an unsaturated organic compound with valence electrons delocalized over covalent bonds in a cyclic molecule.

The chemical reactivity of benzene is also revealing in that it readily undergoes addition reactions and behaves as a nucleophile in substitution reactions (see also the chapters on 'Tea, from Legend to Healthy Obsession' in Part IV and 'A Plant from the East Indies, Camphor' in Part VI).

At ordinary temperatures and elevated pressure, benzene can be hydrogenated in the presence of a finely divided catalyst of platinum to produce cyclohexane.

$$C_6H_6 + 3H_2 = C_6H_{12}$$

Another example of an addition reaction involves halogens such as chlorine, which in the presence of the energy source, ultraviolet light, will yield benzene hexachloride.

The replacement of a hydrogen atom in benzene occurs much more readily than the corresponding replacement in an alkane or cycloalkane. In the presence of concentrated nitric and sulfuric acids at about 50°C, benzene will give nitrobenzene by nucleophilic substitution.

$$C_6H_6 + HNO_3 = C_6H_5NO_2 + H_2O$$

Halogenation of benzene provides another example of nucleophilic substitution. The reaction takes place in the presence of a catalyst, aluminum chloride, when chlorine is passed through the liquid at room temperature.

$$C_6H_6 + Cl_2 = C_6H_5Cl + HCl$$

Chlorobenzene is known to be the result when a measured increase in weight has occurred.

The Friedel-Crafts reaction is an important example of a nucleophilic substitution that provides a very useful synthetic pathway in organic chemistry. Catalyzed by aluminum chloride, benzene can undergo alkylation or acylation to produce an alkyl benzene (such as toluene) or an acyl benzene (such as acetophenone).

$$C_6H_6 + CH_3Cl = C_6H_5.CH_3 + HCl$$
$$C_6H_6 + CH_3.CO.CH_3 = C_6H_5.CO.CH_3 + HCl$$

Reference to the chapters on 'Europe solves a Headache' in Part II and 'Morphine: A Double-edged Sword' in Part V is also advised for more on alkylation.

Finally, mention must also be made of the violent oxidation reaction between benzene vapor and ozone—or indeed between any hydrocarbon and ozone—which ultimately yields carbon dioxide and water. This property of ozone has crucial consequences in the upper atmosphere of the Earth where hydrocarbon pollutants, arising from aircraft at high altitude or from upward diffusion of propellant gases from aerosol cans or refrigeration systems in use in the lower atmosphere, have caused serious reduction in the partial pressure of ozone—commonly referred to as the ozone hole. A key property of ozone is that it absorbs high energy, short wavelength ultraviolet radiation. If ultraviolet radiation is not reduced in intensity to natural levels at the surface of the Earth, it can cause serious damage to plant and animal life.

THEORY OF THE MOLECULAR STRUCTURE OF BENZENE

The structure of benzene with a molecular formula of C_6H_6 has always presented something of a problem. If the molecule were linear, that would suggest that benzene ought to have a similar degree of unsaturation to that of ethyne or acetylene, but benzene is much less reactive in addition and substitution reactions. It is more stable than anyone would expect at first sight. In 1865, Kekule was the first person to suggest a cyclic structure which might be based on a hybrid of resonant forms (Figure 2.7). This idea has been reinforced somewhat by theory in quantum mechanics.

Modern spectroscopic studies of the benzene molecule indicate unequivocally a planar, regularly hexagonal molecule with C-C-C and C-C-H bond angles of 120 degrees. These facts, together with

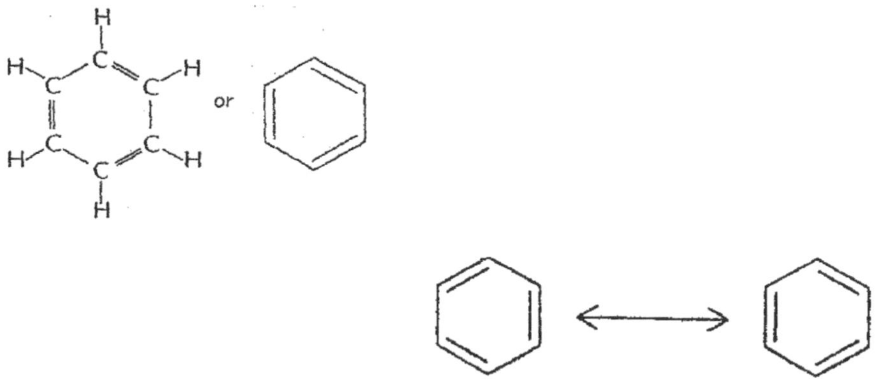

FIGURE 2.7   Kekule's theory of benzene, the structures of resonant hybrids.

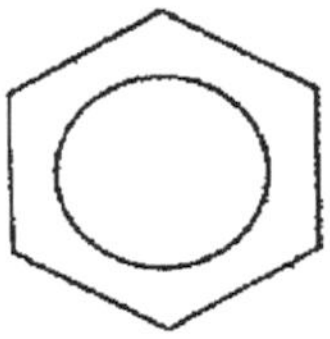

FIGURE 2.8   The delocalized electron structure representing a molecule of benzene.

the relative stability of benzene, compared to alkenes and alkynes, have led to the acceptance of a molecular orbital theory in which the p electron orbitals of neighboring carbon atoms overlap one another thus allowing electrons to pass around the ring above and below the plane of the ring. It is these delocalized electrons which are considered responsible for the aromatic character of benzene and larger compounds containing benzene as a building block. As noted in the chapter entitled 'Tobacco: a profound influence on the World' in Part V, some heterocyclic compounds, such as pyridine, pyrrole and thiophen, are also aromatic for the same reason—the delocalization of electrons around a ring molecule, above and below its plane (Figure 2.8).

While Kekule's structures are not quite consistent with physical and chemical evidence, they are still used today in depicting possible mechanisms of reactions of benzene derivatives, especially when hydrogen atoms in the ring are substituted in the ortho- and para-positions. An example of this behavior is found in the molecule of phenol, which behaves as a weak acid in aqueous solution and reacts much more easily than benzene does in electrophilic substitution reactions involving the 2 and 4 positions (ortho and para) in the carbon ring (see also the chapter on 'Tea, from Legend to Healthy Obsession' in Part IV for more on phenol).

Summary of the characteristics of aromatic compounds

   a) Aromatic compounds burn in air with a luminous, smoky flame, due to the high proportion of carbon in the molecule.
   b) These compounds are somewhat less reactive toward oxidation and reduction than alkenes and alkynes.
   c) They do not undergo addition reactions very readily.
   d) Aromatic compounds behave as weak nucleophiles and are readily involved in substitution reactions with electrophiles.

Questions

1. Give an account of the process of refining petroleum making sure to explain the science involved in each step including fractional distillation and cracking.
2. All manufactured goods have a carbon footprint. Choose an example of a product and explain how the carbon footprint of the product arises from start to end, from the conversion of raw materials to the finished product appearing in the shops.
3. Taking as an example the steroid diosgenin, describe the range of chemical reactions which in principle you would expect to arise from the functional groups present; an ether-like linkage, a carbonyl group, a cyclohexane ring and a degree of unsaturation in the molecule. Describe any limitations in practice.
4. Give a full account of the implications of the high reactivity of introduced hydrocarbons with naturally occurring ozone in the stratosphere of the Earth's atmosphere. Explain what international measures have been undertaken to reduce the release of hydrocarbons to counteract depletion of the ozone layer.
5. Ozone can appear as pollutant at low level in the Earth's atmosphere (or troposphere) in the presence of strong sunlight resulting in a photochemical smog which is injurious to human health. Explain carefully all of the factors involved, especially the chemistry, which give rise to this phenomenon in developed countries of the world.
6. Compare and contrast the physical and chemical properties of cyclohexane and benzene.
7. Draw together the evidence for the delocalized ring structure of benzene molecules and relate this to the Kekule model of resonant hybrids and to modern molecular orbital theory.
8. The simplest polycyclic aromatic hydrocarbon is naphthalene, $C_{10}H_8$, which is a white crystalline solid. It is the main constituent of mothballs. Given the knowledge of benzene, draw a delocalized structure of a molecule of naphthalene and describe the range of chemical properties you would expect naphthalene to have.

SUGGESTED FURTHER READING

N. Applezwei. 1962. *Steroid Drugs*, pp. vii–xi, 9–83. McGraw-Hill.

R. Cooper and G. Nicola. 2014. *Natural Products Chemistry; Sources, Separations and Structures*. CRC Press, Taylor and Francis Group.

C. Djerassi. 2001. *This Man's Pill Reflections on the 50th Birthday of the Pill*, pp. 11–62. Oxford University Press.

J. M. Riddle. 1992. *Contraception and Abortion from the Ancient World to the Renaissance*, p. 28 and references therein. Harvard University Press.

E. W. Straus and A. Strauss. 2006. *Medical Marvels: 100 Greatest Advances in Medicine*, pp. 139–143. Prometheus Books.

REFERENCE

G. Pincus. 1958. *The Hormonal Control of Ovulation and Early Development*. Postgrad Med 24 (6): 654–660.

EUROPE SOLVES A HEADACHE! EMERGENCE OF ASPIRIN

Abstract: *The use of willow bark to relieve symptoms of the ague by riverbank communities led to the eventual development of the great miracle drug, aspirin, by the German pharmaceutical giant, Bayer Company.*

Organic Chemistry

- *carboxylic acids*
- *phenol, a weak acid with a hydroxyl group in a benzene ring*
- *acetylation of the hydroxyl group of salicylic acid to form aspirin.*

Context

- *salicylic acid and aspirin.*

SALICIN AND SALICYCLIC ACID

The active ingredient in willow bark is salicin (Figure 2.9), which is converted naturally within the human body into salicylic acid. Salicin is a glycoside—revealed when it is hydrolyzed to salicylic alcohol and glucose (see the chapters on 'Global Aloe' and 'A Steroid in your Garden' for more about glycosides).

The molecular structure of salicylic acid is shown in Figure 2.10. This molecule is an example of a bi-functional organic compound, whereby the chemical reactivity may be due to either the carboxylic acid group or the phenolic group or both.

CARBOXYLIC ACIDS

Carboxylic acids contain the carboxyl functional group—COOH and examples are shown in **Table 2.4.** Nomenclature, as usual, follows the name of the stem of the corresponding alkane or benzene with the suffix *oic* applied. There are, of course, straight chain carboxylic acids with branched chain isomers, molecules with two or more carboxylic functional groups and aromatic carboxylic acids.

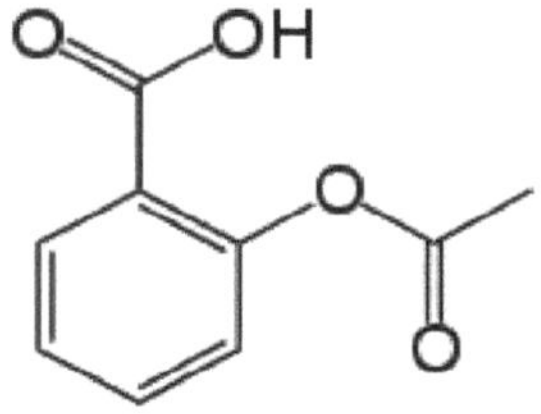

FIGURE 2.9   Chemical structure of salicin.

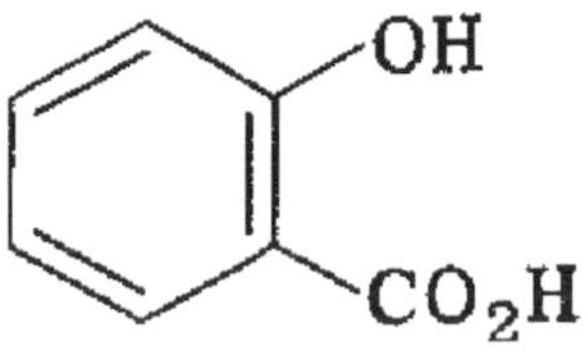

FIGURE 2.10   Molecular structure of salicylic acid.

TABLE 2.4
Examples of Simple Organic Carboxylic Acids

HCOOH	methanoic acid (unsystematic name, formic acid)
$CH_3.COOH$	ethanoic acid (unsystematic name, acetic acid)
$CH_3.CH_2.COOH$	propionic acid
$CH_3.CH_2.CH_2.COOH$	butanoic acid (unsystematic name, butyric acid)
$CH_2Cl.CH(CH_3).COOH$	3-chloro 2-methyl butanoic acid
$C_6H_4(COOH)_2$	benzene dicarboxylic acid (informal name, phthalic acid)

Methanoic acid (b.p. 100°C) and ethanoic acid (b.p. 118°C) are both volatile liquids at ambient temperature and pressure and have pungent smells. The other acids are generally colorless, crystalline solids.

Due to the electronegativity of oxygen atoms relative to those of hydrogen, the hydroxyl bond is polar. This fact accounts for a number of the properties of carboxylic acids—especially those of low molecular mass where the hydrocarbon moiety in the molecule is less influential.

Owing to the influence of hydrogen bonding in the liquid state or in aqueous solution, the lighter aliphatic carboxylic acids are quite soluble in cold water, while aromatic carboxylic acids are only sparingly soluble under the same conditions.

As the name of the class strongly suggests, these compounds

- are acidic in aqueous solution
- form organic salts readily in dilute solutions of inorganic bases
- displace carbon dioxide from inorganic alkali metal carbonates
- form ammonium salts or amides with ammonium hydroxide
- chemically combine with alcohols to form esters (catalyzed in aqueous solution by an inorganic acid).

More on the formation, properties and uses of esters is to be found in the chapter 'A Steroid in Your Garden' in Part II.

PHENOL AND PHENOLIC COMPOUNDS

Phenol (Figure 2.11) consists of a benzene ring in which one of the hydrogen atoms has been replaced by a hydroxyl group. More on the properties and uses of phenol and phenolic compounds is to be found in the chapter, 'Tea: From Legend to Healthy Obsession!' in Part IV.

Phenol dissolves sparingly in water since the polar hydroxyl group is able to form hydrogen bonds with water molecules. More importantly, the hydroxyl group in phenol partially dissociates in aqueous solution to form a phenoxide anion and a hydrogen cation, which are solvated because water is a polar solvent. The phenate anion is also stabilized by delocalization of electrons over the benzene ring.

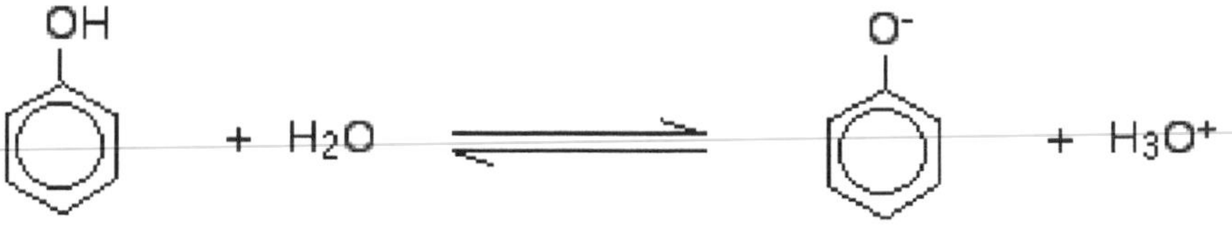

FIGURE 2.11 Partial dissociation of phenol in aqueous solution involving ions and the molecule in equilibrium.

The dissociation is quite reversible at room temperature. In consequence, molecules and ions are in dynamic equilibrium and aqueous solutions of phenol have the properties of a weak acid:

- reacting with strong bases to form a salt and water
- unable to react with weaker bases such as sodium carbonate.

Even though phenol is different from aliphatic alcohols in many ways, phenol will form esters with a carboxylic acid. As carboxylic acids are weak acids themselves, the reaction is slow but may be sped up acceptably by using either an acyl chloride, such as ethanoyl chloride or alternatively and more safely, ethanoic anhydride. Phenyl ethanoate would be the product.

$$CH_3.COCl + C_6H_5.OH = CH_3.COO.C_6H_5 + HCl$$

Phenol is an important building block found in many natural products. Ease of electrophilic substitution (see Glossary) in the ortho and/or para positions (also known as the 2 and 4 carbon positions) of the aromatic ring is a notable chemical property of phenol. The directing effect of the hydroxyl group is so strong in phenol that during a chemical preparation it is often difficult to control the reaction to just mono-substitution.

More on the properties and uses of phenol and phenolic compounds is to be found in *'Tea: From Legend to Healthy Obsession!'* in Part IV and in *'Aromatic Herbs'* in Part III.

ACETYLATION OF THE HYDROXYL GROUP OF SALICYLIC ACID TO FORM ASPIRIN

Phenol undergoes another type of reaction in anhydrous conditions which involves replacement of the hydrogen atom of the hydroxyl group by a carboxyl group or by an ether linkage or by an acetyl group.

Acetylsalicylic acid, commonly known as aspirin, may be prepared in the laboratory by substitution of the hydrogen atom of the phenol group of salicylic acid with an acetyl group **(Figure 2.12)**. Substitution of the phenol group may be achieved under anhydrous conditions using an acyl chloride, ethanoyl chloride or preferably by using an acid anhydride, ethanoic anhydride, as in the reaction shown in the following figure, since a hazardous by-product, gaseous hydrogen chloride, is avoided. Further reference should also be made to the chapter entitled 'Morphine: A Double-Edged Sword' in Part V where the acetylation reaction is examined closely.

Aspirin is a colorless crystalline solid (melting point 135°C) that is soluble in water. The drug is well known as an analgesic (pain killer) and as a pyretic (reduces fever or body temperature). Besides aspirin, another derivative of salicylic acid deserves mention. Methyl salicylate, oil of wintergreen, occurs in many plants. Owing to its fragrant smell, methyl salicylate is used in perfumery.

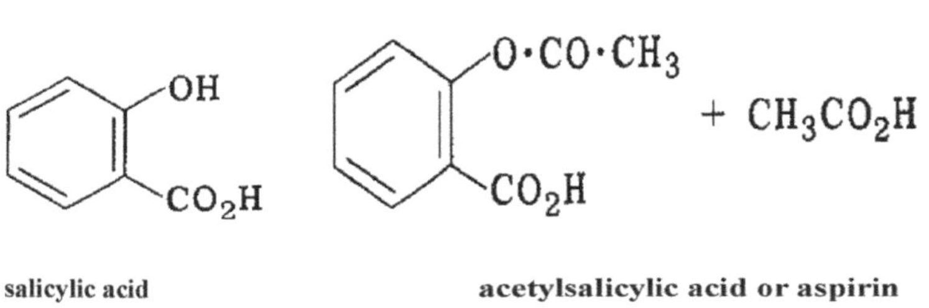

salicylic acid   acetylsalicylic acid or aspirin

FIGURE 2.12   Salicylic acid and acetylsalicylic acid commonly known as aspirin.

HISTORICAL NOTE

From the earliest times, it was known that the chewing of the bark of the willow tree, Salix alba, reduced fever and inflammation. The ancient Greek physician, Hippocrates, is recorded as having recognized its therapeutic benefits.

In 1763, the Reverend Edmund Stone of Chipping Norton, Oxfordshire, England read an obscure paper to the Philosophical Society of London entitled 'An Account of the Success of the Bark of the Willow in the Cure of Agues'. Later, he wrote about this lecture in a letter to the Right Honorable George, Earl of Macclesfield, who was the President of the Royal Society at the time.*

In the 1800s, pharmacists created salicylic acid in its acetylated form (acetylsalicylic acid)—more commonly known by its brand name, Aspirin. Aspirin was first isolated and synthesized by Felix Hoffmann, a chemist with the German company Bayer and marketed in 1897. The most widely used drug in the world continues to be Aspirin, which remarkably has remained essentially unchanged for over 2,500 years.

Questions

1. Give the systematic names and molecular structures of the isomers of pentanoic acid.
2. Name the three isomers of benzene dicarboxylic acid and give their structural formulae.
3. Give the systematic names of oxalic acid, $(COOH)_2$; lactic acid, $CH_3.CH(OH).COOH$; tartaric acid, $[CH(OH).COOH]_2$; benzoic acid, $C_6H_5.COOH$ and cinnamic acid, $C_6H_5.CH:CH.COOH$.
4. Explain the origin of hydrogen bonding in carboxylic acids and the nature of dimers in glacial ethanoic acid. Account in turn for the differences in the physical properties of low molecular weight carboxylic acids, high molecular weight carboxylic (fatty) acids and aromatic carboxylic acids.
5. Knowing the properties of a carboxylic acid and phenol, describe in full the chemistry of salicylic acid given that it is an example of a bi-functional organic compound.
6. Give examples of plants which are sources of oil of wintergreen and explain how it is extracted and used commercially.
7. In phenol, the *ortho* and *para* positions in the carbon ring have been revealed in this chapter. Identify the *meta* position in the ring.
8. Explain why the OH group in phenol has such a strong directing effect within the carbon ring in reactions involving electrophiles.
9. Explain why phenol, C6H5OH and benzyl alcohol, C6H5CH2OH, are so different even though both molecules contain a phenyl group and an alcohol group.

SUGGESTED FURTHER READING

D. Jeffreys. 2004. *Aspirin, Story of a Wonder Drug*. Bloomsbury.
C. C. Mann and M. L. Plummer. 1991. *Aspirin Wars*. A. A. Knopf.
E. W. Straus and A. Strauss. 2006. *Medical Marvels: 100 Greatest Advances in Medicine*. Prometheus Books.

REFERENCE

* *'Philosophical Transactions' 1683–1775*. 1763. 53: 195–200. Published by the Royal Society.

ATTACKING MALARIA: A SOUTH AMERICAN TREASURE AND A CHINESE MIRACLE

Abstract: *The huge cultural and economic impact of an extract from cinchona bark upon European colonization of tropical and sub-tropical regions of the world can scarcely be exaggerated.*

Spanish colonists learned about cinchona from the native South Americans, brought the plant back to Europe and as a consequence reduced the spread of malaria. Later, scientists discovered how to isolate and refine the active chemical ingredient, now known by the name quinine.

In recent times, initially secret research in China led to the development of a new anti-malarial drug called artemisinin.

Organic Chemistry

- *carbon—oxygen bonds in organic compounds*
- *oxygen—oxygen bonds in organic compounds*
- *peroxides*
- *chemiluminescence.*

Context

- *quinine*
- *artemisinin.*

MALARIA

Malaria is a vector-borne disease transmitted by the bite of a female mosquito infected with single-celled (protozoan) parasites (known as plasmodium). The tiny parasites pass through the bloodstream of the human victim and travel to the liver where they mature and reproduce before affecting the whole body by attacking red blood cells. Symptoms of malaria typically include headache, feverishness and fatigue. If not treated, malaria can cause death. The areas in the world most associated with malaria are shown in red in Figure 2.13.

Two major drugs are employed in the fight against malaria and both originate directly from natural products; quinine from the bark of the cinchona tree found in South America; artemisinin from the leaves of *Artemisia annua* native to China. The latter discovery is particularly important as the

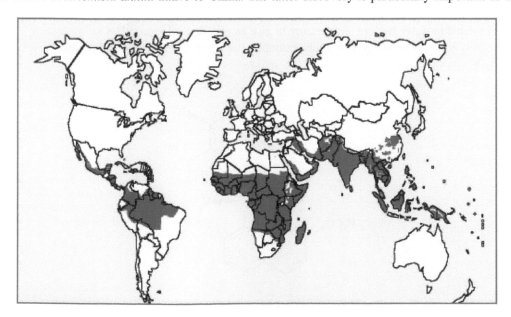

FIGURE 2.13   Parts of the world where malaria is endemic are shown in red on the map.

effectiveness of drugs based solely on quinine has gradually diminished as the infecting parasites have developed resistance to the quinine-based drugs. Subsequently, artemisinin has become the treatment of choice for malaria. However, the World Health Organization (WHO) called for cessation in the use of single use of artemisinin preparations in 2006 in favor of combinations of artemisinin together with another malaria drug in order to reduce the risk of the parasites developing resistance. Thus, artemisinin is usually combined with a synthetic derivative of quinine, known as chloroquinine. In this dual dose, the drugs reinforce one another in addressing malaria in that they have complementary roles; the former is quick acting whilst the latter reduces inflammation. However it remains to be seen whether the strategy of combination therapy will be entirely successful in the management of malaria.

CINCHONA

Cinchona is an evergreen shrub or small tree indigenous to the high Andes of South America.

The botanical name of the genus Cinchona was given by Linnaeus in 1742 from the Indian name, *Quinaquina*, derived from the Quechua language of Ecuador, Peru and Bolivia. The Quechua people used an extract from the bark as a relaxant for muscles to combat extreme cold in the high mountains. It was also noticed that the extract had a favorable impact on malaria. Later use of an extract from the bark of the cinchona tree led to significant reduction in malaria among the ranks of Conquistador invaders and helped strengthen the might of Spain for generations. Because of its medicinal property, Spanish colonists brought the shrub back to Europe around the early 1600s. Later research expeditions left from France, the Netherlands and Great Britain. Eventually, the Dutch and British cultivated cinchona for this mysterious extract in the East Indies and in the Indian sub-continent respectively.

ISOLATION OF QUININE

This substance was first isolated in 1820 by Pierre Pelletier from the bark of the cinchona shrub. He separated a yellow gum with a bitter taste which he called quinine. However, the exact chemical structure presented in **Figure 2.14** was not fully elucidated until much later. Even today, quinine is still recognized as one of the most effective drugs in the treatment of malaria although details of the mechanism of its disruption of the life cycle of the plasmodium parasite are not understood.

The quinine molecule has two distinctive parts; an aromatic component and an amine component. As a consequence, quinine

- is soluble in water at room temperature to give an alkaline solution (pH 8.8)
- readily forms salts with strong acids, examples being quinine sulphate and quinine hydrochloride.

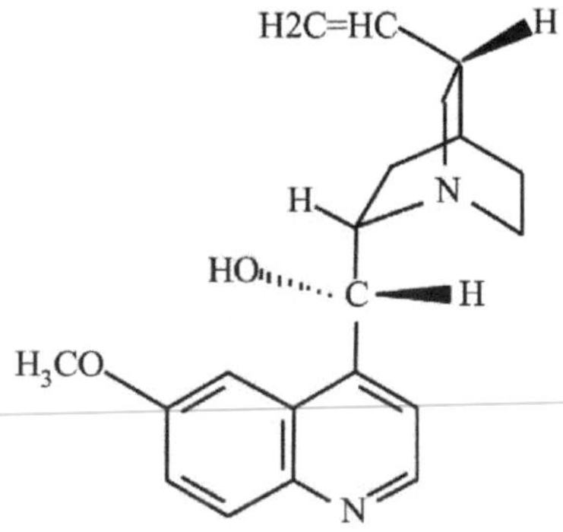

FIGURE 2.14   The structure of quinine, the anti-malarial drug from the cinchona bark.

SYNTHESIS OF QUININE IN THE LABORATORY

The famous and brilliant British chemist, William Perkin, tried to synthesize quinine in 1856 but was unsuccessful. Eventually, quinine was synthesized in 1944. Quinine was the first drug from a natural product to be synthesized although the route was so complex it could not be made a commercial success. Consequently, quinine is still obtained directly from the cinchona bark. Nowadays, quinine is then chemically modified synthetically to make quinine derivatives that are more powerful as anti-malarial drugs.

However, it should be remembered that Perkin's persistence in the early research of 1856 led remarkably to the discovery and production of a mauve-colored compound, which he established as a derivative of aniline. Inadvertently, he had fashioned the world's first synthetic dye, which became the basis of the entire aniline dye industry. For more on dyes, see the chapter on 'Woad (Isatis tinctoria) and Indigo' in Part VII.

USES OF QUININE

Beyond its application as an anti-malarial drug, quinine is widely utilized in the modern drinks industry. By repute, British administrators and army personnel working for the colonial service in the Indian sub-continent during the 19th century dissolved their bitter-tasting anti-malarial treatment or tonic (quinine) in gin (a solution of ethanol in water). Henceforward, gin and tonic became established as a fashionable and well-liked alcoholic drink. In the English speaking world, tonic water is appreciated as a mixer for cocktails, especially those based on gin or vodka.

To this day, the bitter taste of quinine is used in dilute solution to produce what is known and sold as 'tonic water'. Quinine, or a quinine salt such as quinine hydrochloride, is dissolved in small quantity in citrus drinks in order to enhance the flavor of soft drinks such as bitter lemon and bitter lime.

ARTEMISININ

More recently, a new and completely different anti-malarial miracle drug, artemisinin, has been extracted from the leaves of *Artemisia annua* (Annual Wormwood) in China.

HISTORICAL NOTE

The plant Artemisia annua has been used by Chinese herbalists for over two millennia. An extract was believed to have been used in the treatment of skin diseases and malaria. The anti-malarial property of the extract was first specifically described in the 4th century within the classic Chinese text 'The Handbook of Prescriptions for Emergencies'.

In the 1960s, a research program was set up by the Chinese army to find an adequate treatment for malaria. By 1972, artemisinin had been discovered in the leaves of Artemisia annua. Screening of over 5,000 traditional Chinese medicines revealed that artemisinin was the most effective drug in dealing with malaria parasites in a patient. Initially, due to the secret nature of the research program, the work was never given the full international recognition it deserved, and yet today the world is a grateful and beneficial recipient of chemical derivatives based on artemisinin.

Artemisinin (Figure 2.15) has a complex structure including a 6-membered lactone ring (for more on lactones see the chapter entitled 'A Steroid in Your Garden'. The structure also contains an unusual peroxide linkage, which is believed to be involved in the anti-malarial effectiveness of the drug and may also account for its relatively rapid medical action compared to quinine. The World Health Organization (WHO) recognizes that artemisinin is very effective in the prevention and treatment of malaria even in cases where the parasite responsible is resistant to quinine. To avoid the

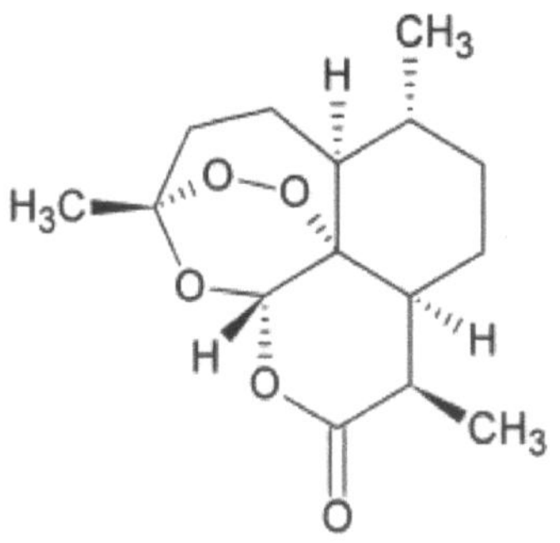

FIGURE 2.15   Molecular structure of artemisinin.

possibility of drug resistance to artemisin treatment, WHO recommends combination therapy of artemisin and quinine derivatives respectively.

CARBON–OXYGEN SINGLE BONDS AND OXYGEN–OXYGEN SINGLE BONDS IN ORGANIC COMPOUNDS

The electronegativity of carbon and oxygen atoms is similar arising from the fact that they are close to each other in atomic structure and position in the Periodic Table. Consequently, carbon–oxygen single bonds are covalent and strong with little charge separation or dipole effect.

The peroxide link formed by a single covalent bond between two oxygen atoms, O–O, is relatively weak by comparison with a single C–O bond. We all know that chains, however long, are only as strong as the weakest link because that is the point where a break is most likely to occur. Physically and chemically, the molecules of peroxide compounds are unstable, since they are likely to break at the weakest point, the single O–O bond, resulting in the release of free radicals (see Glossary), which are highly reactive and destructive to tissues in the human body.

The instability of organic peroxide compounds may be contrasted with the great stability of ethers where the strength of the single C–O–C covalent bonds results in a family of substances that is relatively stable with little chemistry. These compounds are often volatile liquids at ambient temperature and pressure due to an absence of hydrogen bonding between adjacent molecules. For more on ethers, see the chapter on 'Maca from the High Andes in South America' in Part IV.

Furthermore, the stability of C–O single bonds should not be confused with the reactivity of C–O double bonds, which are somewhat strained physically and weakly dipolar, leading to the extensive chemistry of the large families of organic compounds which contain the C = O double bond; namely, aldehydes, carboxylic acids, esters and ketones. This book provides many references to the diverse chemistry of each of these families of organic compounds, which are covered extensively in the chapters on 'Asian Staple—Rice', 'A Plant from the East Indies, Camphor', 'A Steroid in Your Garden', 'Morphine—A Double-Edged Sword' and 'Europe Solves a Headache'.

THE PEROXIDE LINK AND THE PEROXIDE BRIDGE IN ARTEMISININ

A peroxide is a compound containing the link of two oxygen atoms joined together by a single covalent bond as in hydrogen peroxide—H_2O_2—or in a generalized organic peroxide, R'.O.O.R". The peroxide bond is weak, which renders peroxide compounds unstable, hence they readily decompose to form highly reactive free radicals, R'O and R"O. Due to this instability, hydrogen peroxide and organic peroxides are found in strictly limited quantities in the natural environment.

Owing to their property as a bleaching agent, peroxide compounds are used in detergents and in cosmetic colorant treatments for human hair. In the laboratory and in some industrial pharmaceutical processes, organic peroxides are utilized to good effect as intermediates or building blocks in step-wise and lengthy preparations.

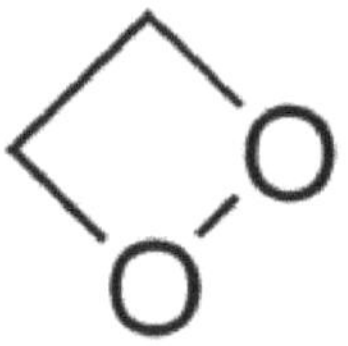

FIGURE 2.16   1, 2-Dioxetane also called 1, 2-dioxacyclobutane.

It is interesting to note that the firefly produces a small amount of the peroxide, 1, 2-dioxetane (Figure 2.16), which decays spontaneously to produce electronically excited molecules of acetaldehyde. As each molecule returns to the ground electronic state, a quantum of visible light is released—an instance of chemi-luminescence. Quaint glow sticks produced for human entertainment have 1, 2-dioxethanedione, C_2O_4, embedded within them, which slowly breaks down in a similar way to yield carbon dioxide and emission of light.

A more sobering consideration is the care required in the storage of liquid organic compounds in the laboratory—particularly ethers. When ether is stored in a Winchester bottle (see Glossary) with some air inevitably enclosed and in the presence of light, unstable peroxides will form very slowly over a period of time measured in months, rendering the bottle somewhat explosive and hazardous to use. This effect is counteracted simply by keeping ether over potassium hydroxide, which destroys the peroxide impurity as it develops.

Questions

1. Explain the chemical properties of quinine by reference to its molecular structure.
2. Give an account of the behavior of the peroxide link seen in artemisinin and relate this to the properties of free radicals in organic chemistry.
3. Explain why peroxides are so unstable.
4. Potassium hydroxide is added to ethers when they are stored in order to avoid peroxide formation. Can you explain how this precaution avoids the risk of accumulation of dangerously explosive peroxides?
5. Give an account of the different chemistry of each of the three C-O bonds in artemisinin.
6. Give a full account of the implications of the high reactivity of introduced hydrocarbons with naturally occurring ozone in the stratosphere of the Earth's atmosphere. Explain what international measures have been undertaken to reduce the release of hydrocarbons to counteract depletion of the ozone layer.
7. Ozone can appear as a pollutant at low levels in the Earth's atmosphere (or troposphere) in the presence of strong sunlight resulting in a photochemical smog, which is injurious to human health. Explain carefully the chemistry that gives rise to this phenomenon in developed countries of the world.

SUGGESTED FURTHER READING

R. Cooper and G. Nicola. 2014. *Natural Products Chemistry; Sources, Separations and Structures*. CRC Press, Taylor and Francis Group.
H. Hobhouse. 2005. *Seeds of Change. Six Plants That Transformed Mankind*, pp. 3–50. Counterpoint.

REFERENCES

J. Jaramillo-Arango. 1949. A *Critical Review of the Basic Facts in the History of Cinchona*. J Linn Soc Bot 351: 272–309.
R. B. Woodward and W. E. Doering. 1944. *The Total Synthesis of Quinine*. J Am Chem Soc 66 (5): 849.

A STEROID IN YOUR GARDEN

Abstract: There is a very important medicine growing in your garden! In western medicine, the drug known as digitalis has been exploited for its medicinal qualities for many years. Digitalis is obtained from the roots and seeds of the foxglove. In fact, the foxglove contains several important but highly toxic, chemically related steroidal glycosides.

Organic Chemistry

- *the hydroxyl functional group in alcohols (primary, secondary, tertiary)*
- *the carboxy functional group in esters.*

Context

- *cyclic esters—known as lactones*
- *lactones as building blocks in nature.*

THE FOXGLOVE AND DIGOXIGENIN

The foxglove (Latin name*: Digitalis purpurea*) is a wild plant, a native of the woodland margin in temperate climes, which is often cultivated in domestic gardens for its ornamental value.

Foxglove, including the roots and seeds, contains chemicals known as cardiac glycosides. One of these chemicals has the common name digitalis or its chemical name, digoxigenin **(Figure 2.18)**. This chemical is composed of two distinctive building blocks, which occur frequently in the natural world; the carbon skeleton of a steroid (see the chapter on 'Central America's Humble Potato!' in Part II) and that of the five-membered cyclic ester (RCOOR) known as a lactone (see **Figures 2.18, 2.19** and **2.20** for examples).

FIGURE 2.17   Foxglove (*Digitalis purpurea*).

   Digitalis is an example of a drug derived from a plant used by folklorists and herbalists although it is difficult today to determine what amounts of active drug were present in those early herbal preparations. An extract of digitalis was used for the first time in 1785 as the modern era of therapeutic science was beginning. William Withering, a Fellow of the Royal Society, was a physician active at the hospital in Birmingham, England. He published 'An Account of the Foxglove and some of its Medicinal Uses', which contained notes on the medical effects of digitalis on congestive heart failure, characterized by low heart output. He had learned of the folk remedies used by people in Shropshire, UK where he had grown up. Digitalis is used to control a slow heart rate and to strengthen the contraction of heart muscles, thereby producing a stronger pulse. However, digitalis is very toxic and must be administered carefully. An overdose can easily be fatal. It can still be prescribed for those patients who suffer from atrial fibrillation (an erratic heartbeat), especially if they also have congestive heart problems.

   A modern application of digoxigenin is in the field of analysis called immuno-histochemical staining (IHS). The technique is widely employed in molecular biology to reveal visually the

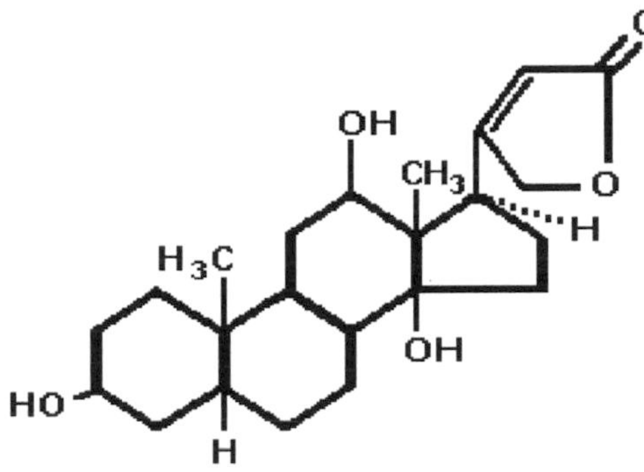

FIGURE 2.18   Molecular structure of digoxigenin.

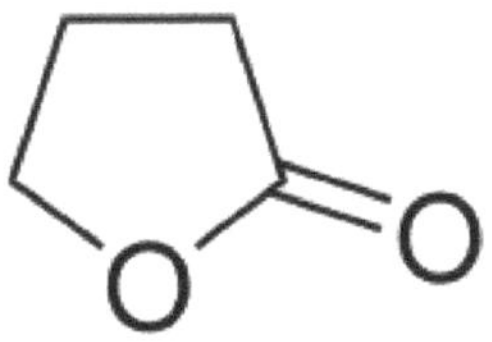

FIGURE 2.19   The structure of 4-hydroxybutyric acid lactone, also known as γ-hydroxybutyric acid lactone.

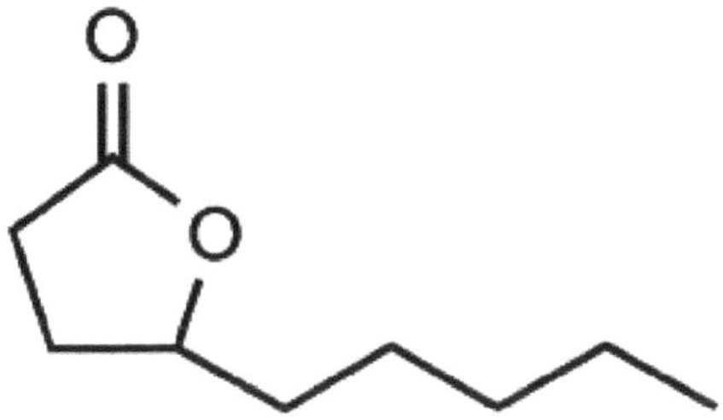

FIGURE 2.20   Chemical structure of gamma-nonalactone.

presence of antigens causing abnormality in cells, as in cancerous tumors and was first reported in the scientific literature by Coons et al. in 1941. IHS has proved to be an excellent means of detecting the protein belonging to an antigen within body tissue—especially in neuroscience where tumors may be located precisely within nerve tissue and brain cells.

Esters are Formed from Alcohols and Acids

The alcohols are a class of organic substances forming an homologous series with the general formula $C_nH_{2n+1}OH$ containing the characteristic functional group: the hydroxyl. They are rarely found in a free state in nature although a little may be present in over ripe fruit. The individual names of alcohols are derived by adding the suffix, ol, to the name of the corresponding alkane, thus: methanol, CH_3OH; ethanol, C_2H_5OH; propanol, C_3H_7OH etc. Where isomerism occurs in propanol and in succeeding members of the series, the position of the hydroxyl group in the longest carbon chain is indicated by inserting a number before the 'ol'.

Thus, there are three types of alcohol; primary, secondary and tertiary depending on the chemical environment of the carbon atom to which the hydroxyl is bonded.

Primary	R'.CH$_2$OH
Secondary	R'R".CHOH
Tertiary	R'R"R"'COH

There is a marked difference between alkanes and alcohols with regard to boiling point and solubility in water. The difference arises unambiguously from the ability of hydroxyl groups to form hydrogen bonds by electrostatic attraction due to their dipolar nature. This molecular association has the effect of increasing molecular weight. Even the alcohols of lowest molecular weight are colorless liquids at normal temperature and pressure. As the alkyl moiety increases in proportion in the molecules of larger members of the series so the influence of the hydroxyl group declines and physical properties relate more closely to those of the alkanes.

Alcohols undergo two types of chemical reaction directly arising from the hydroxyl group which may break at the hydrogen-oxygen bond or at the carbon-oxygen bond.

Dehydration of ethanol occurs when the vapor is passed slowly at 200°C over a finely divided catalyst in aluminum oxide. Diethyl ether ($CH_3CH_2.O.CH_2CH_3$) is the product formed from the breaking of the hydroxyl bond.

$$2C_2H_5OH = CH_3.CH_2.O.CH_2.CH_3 + H_2O$$

However, at a higher temperature of 300°C, ethene is formed as the stronger hydroxyl bond is removed.

$$C_2H_5OH = CH_2CH_2 + H_2O$$

This reaction provides a renewable, synthetic pathway in the chemical industry to alkenes and polymers from a feedstock of ethanol produced by the fermentation of glucose obtained from commercially grown plants such as sugar cane.

A primary alcohol may be oxidized through an aldehyde as an intermediate to form a carboxylic acid.

$$R.CH_2.OH + 'O' = R.CHO + 'O' = R.COOH$$

A typical oxidizing agent for the reaction, which is performed under reflux conditions, is acidified potassium dichromate solution.

Esters are formed by the elimination of water from a reaction between an alcohol and a carboxylic acid, for example:

$$C_3H_7OH + CH_3.CO.OH = CH_3.CO.OC_3H_7 + H_2O.$$

The reaction is effected by gently refluxing the reagents in the presence of a small quantity of a mineral acid, sulfuric or hydrochloric, as a catalyst.

PROPERTIES AND USES OF ESTERS

Esters are formed from alcohols and carboxylic acids, whose properties are fully described in the chapter entitled 'Europe Solves a Headache!' and reinforced in 'Morphine: A Double-Edged Sword'.

The protocol for naming esters parallels that for the naming of salts in inorganic chemistry. Since esters are made from an alcohol and an acid, the alkyl group from the alcohol comes first and the acid stem second.

While esters can be made synthetically in the laboratory, esters occur naturally in many different flowers and fruits. Indeed, ethyl ethanoate smells strongly of the confection known as pear drops.

Esters have a sweet, fruity aroma that serves well for applications in the perfume industry. Esters are also used as a flavoring agent in the processed food and soft drinks industries.

Due to the presence of the carboxyl group, esters tend to be polar molecules so even those of low molecular weight are liquids. These esters find application as a solvent for printing ink where their characteristic smell is noticeable when marker pens are applied by a reader to highlight text on paper.

Esters are also applied in the cosmetics industry where they are used as the medium in which the active ingredients of creams and soothing or healing salve are dissolved or suspended. Furthermore, esters contribute to the aesthetic feel sof the product on the skin.

INTRODUCING CYCLIC ESTERS KNOWN AS LACTONES

The name of the class, lactone, arises from the intra-molecular dehydration of lactic acid, $CH_3CH(OH).COOH$. The Latin name for sour milk is *lactis* owing to the presence of lactic acid. A lactone (Figure 2.19) is a cyclic ester that can be formed from an intra-molecular reaction as the condensation product of an alcohol group –OH and a carboxylic acid group—COOH. Lactones are characterized by a closed ring consisting of two or more carbon atoms and a single oxygen atom with one of the carbon atoms being part of a ketone functional group, typical of an ester (see the top right of the structure of digoxigenin presented earlier in this chapter).

Individual lactones are named according to the precursor carboxylic acid;

- aceto two carbon atoms
- propio three carbon atoms
- butyro four carbon atoms
- valero five carbon atoms
- capro six carbon atoms.

The nomenclature of lactones involves one further refinement. The first carbon atom along the chain of the parent molecule from the carbon atom in the COOH group is labeled alpha (α), the second beta (β) and so on.

Lactone molecules with three- or four-membered rings are physically strained and as a consequence are quite reactive. In contrast, lactones with five- or six-member rings are relatively straightforward to prepare and are much more stable.

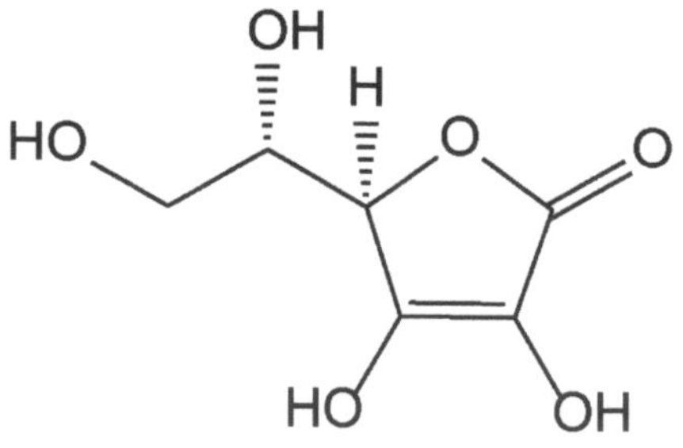

FIGURE 2.21   The structure of Vitamin C, ascorbic acid.

LACTONES AS BUILDING BLOCKS IN NATURE

Lactone rings occur widely in nature as building blocks within larger molecules which form a part of neurotransmitters or of various enzymes or of ascorbic acid also known as vitamin C (Figure 2.21).

VITAMINS

The term vitamin is derived from 'vitamine'—a compound word formed from vital and amine.

Currently, there are thirteen recognized vitamins: vitamins A to E—including a range of B vitamins—and vitamin K.

Vitamins fall into two broad categories. The fat soluble vitamins, vitamins A, D, E and K, can be stored by our bodies in the liver or in fatty tissues. They are stored until they are required, which consequently means they generally do not need to be ingested as frequently. Water soluble vitamins, on the other hand, are not stored in the body. As such, they must be a regular part of the diet in order to avoid deficiency.

Vitamins are vital for good human health; they are an important part of our diet. They perform a range of roles in the body. For example, a number of the B vitamins are important for making red blood cells and in the metabolism of a variety of compounds during digestion. Others have uses in more specific parts of the body; for example, vitamin A is important for good eyesight, whilst vitamin K plays a major role in the clotting of blood. Conversely, deficiencies of vitamins can also have effects; a lack of vitamin C can lead to scurvy, the bane of sailors before the role of vitamin C was understood. A lack of vitamin K can cause bleeding problems which is why newborn babies are given a shot containing the vitamin to prevent potential brain damage.

VITAMIN C, ASCORBIC ACID

Vitamin C is an essential nutrient for sound human health. Vitamin C is a primary metabolite that is directly involved in normal growth, development and reproduction. Also, it plays a role as a redox

HISTORICAL NOTES ON VITAMIN C

A work published in 1753 suggested that citrus fruits (limes) contained certain compounds that could treat scurvy. Scurvy became an endemic disease between the 17th and 19th centuries because of insufficient dietary intake of fresh fruit and vegetables. Today, we know that vitamin C has the ability to cure and prevent scurvy.

The discovery of vitamin C began in the late 16th century when French explorers were saved from effects of scurvy by drinking a tea made from the arbor tree during long sea voyages. Later, it was noted that lemon juice could prevent people from getting scurvy. By 1734, it had been concluded that people who did not eat fresh vegetables and greens would get the disease—a clear risk for sailors denied access due to a long sea voyage in the days of sail. All seamen were thus provided with citrus fruits. The famous British naval captain, James Cook, routinely supplied his men with limes during long voyages of exploration and hydrographic survey in the late 18th century. In fact, the practice was commonplace in the Royal Navy of the time so much so that the American term for the British, 'limeys', arises from it. These observations led to important breakthroughs in the understanding of scurvy through experiments involving guinea pigs and became one of the first examples of the use of animal models to study disease. The first chemists to isolate vitamin C, ascorbic acid, were Svirbely and Szent-Gyorgyi for which they received the Nobel Prize for Medicine in 1937. The approach takes advantage of the acidic functional groups present in the molecule since ion exchange resins are used to remove the acidic cations of vitamin C from aqueous solution. When a dilute solution of a strong acid is eluted through the resin as a second step, the anions of the strong acid are retained and vitamin C or ascorbic acid is released.

The first synthesis of vitamin C was achieved by Haworth and Hirst and also resulted in the award of the Nobel Prize for Chemistry in 1937. Mass production of vitamin C by the Swiss pharmaceutical giant, Hoffmann-La Roche, came twenty years later.

agent and catalyst in a broad array of biochemical reactions and processes. Vitamin C acts as a reducing agent donating electrons to various enzymatic and non-enzymatic reactions.

Vitamin C is found in fresh vegetables and fruit and is also present in animal organs such as the liver, kidney, and brain. Due to its antioxidant properties, vitamin C has been widely used as a food additive to prevent or limit oxidation.

COMMERCIAL USES OF LACTONES

As is the case with chain or branched chain esters, some lactones, since they are cyclic esters, are used for flavoring processed foods and drinks and as fragrances particularly when specific aromas are required. For instance, gamma nonalactone, presented in Figure 2.20, smells of coconut.

Artemisinin, a complex lactone, is employed in the prevention and treatment of malaria—see the chapter on 'Attacking Malaria: a South American Treasure and a Chinese Miracle' in Part II.

Questions

1. Name the following alcohols; CH_3CH $(OH).CH_3$; $CH_3.CH_2.CH_2.CH_2OH$; $CH_3.CH_3.$ $CH.CH_2OH$; $CH_3.CH$ $(OH).CH_2.CH_3$; $CH_3.CH_3.CH_3.COH$. Identify the primary, secondary and tertiary alcohols.
2. Give an account of the uses of different alcohols—giving emphasis to ethanol—and their value as feedstock for a variety of industrial processes.
3. Compare and contrast the oxidation of primary, secondary and tertiary alcohols.
4. Given knowledge of the chemistry of esters, describe and give examples of the kinds of reactions lactones will undergo.
5. Three and four member ring lactones are quite reactive, whereas lactones with five- or six-member rings are much more stable. What is the reason for this in physical terms?
6. What is a vitamin?

SUGGESTED FURTHER READING

G. F. Ball. 2004. *Vitamins: Their Role in the Human Body.* John Wiley and Sons.

M. B. Davies, J Austin, and D. A. Partridge. 1991. *Vitamin C: Its Chemistry and Biochemistry.* Royal Society of Chemistry, Paperback Series.

R. B. Rucker, J. Zempleni, J. Suttie, and D. McCormick, Eds. 2007. *Handbook of Vitamins*, 4th Edition. Taylor and Francis.

REFERENCE

A. H. Coons, H. J. Creech, and R. N. Jones. 1941. *Immunological Properties of an Antibody Containing a Fluoroscein Group.* Proc Soc Exp Bio Med 47: 200–202.

AFRICA'S GIFT TO THE WORLD

Abstract: The Madagascan periwinkle becomes Africa's great gift to the world. A chain of serendipitous events led the Eli Lily Pharmaceutical Company, based in the USA, to isolate the drug, vincristine, which is still used today in the fight against childhood leukemia.

Organic Chemistry

- *aromatic chemistry with a nitrogen atom within an organic ring, i.e. an indole, and alkaloids*
- *isolation of chemicals from plants by fractional distillation and acid-base extraction.*

Context

- *vincristine and vinblastine*
- *alkaloids and indoles.*

Discovery of the Periwinkle Plant and Its Properties

Periwinkle, also known as *Catharanthus rosea*, is a tropical perennial often grown as an annual in temperate climates. The African plant was found only in Madagascar (Figure 2.22).

It was first described in the mid-18th century. In 1757, the Madagascar periwinkle was brought to Europe as an ornamental plant and was cultivated in European gardens. By the late 18th century, widespread distribution throughout the tropics led to its adoption as a medicinal plant wherever it became established:

- in Cuba and in Puerto Rico an infusion of flowers together with a few drops of alcohol (ethanol) added was used as an eyewash for infants
- in Latin America, the leaf tea has been used as a gargle for sore throat and laryngitis
- in India, the fresh juice squeezed from the leaves was used for wasp stings

FIGURE 2.22   Location of Madagascar in the Indian Ocean off the east coast of southern Africa.

- in Vietnam, herbalists use the leaf and stem tea as a treatment for everything from menstrual difficulties to malaria
- in Asian cultures, South Africa, and Caribbean islands, periwinkle tea was useful as a folk cure for diabetes.

In the early 20th century, patent medicines containing periwinkle were touted as a cure for diabetes. 'Vinculin' was sold in Great Britain while 'Convinca' was sold in South Africa but both were later shown to have no genuine medical influence in reducing levels of blood sugar. Despite this, consumption of the Madagascar periwinkle as a spurious remedy for diabetes continued and, remarkably, led scientists to use the plant as the source for the development of an anti-leukemia drug.

Indole is an important building block for many naturally occurring alkaloids, which include significant, complex chemical compounds extracted from the Madagascan periwinkle. Both indole and the indoles as a family of compounds are described in more detail in chapters entitled 'Maca from the High Andes in South America' in Part IV and 'Woad and Indigo' in Part VII.

SERENDIPITY

Probably more drugs have been developed through a serendipitous approach by scientists than through planned attack. Educated observation of abnormal events can be further investigated. However, the discovery of the exquisitely complex chemicals from the Madagascar periwinkle was serendipitous. Rather than an inspirational, Archimedean moment of "Eureka", the two active, naturally occurring chemicals extracted from the Madagascar periwinkle were developed from research and careful observation. These compounds were eventually made commercially and sold as drugs. They were called Velban and Oncovin and were developed by a US pharmaceutical company based in Indiana: the Eli Lily Company.

Traditional Madagascan healers used the periwinkle for treating diabetes, which led western scientists to collect samples of the plant and subsequently to study the plant extract further. Eventually, using sophisticated chemical and biological techniques and some luck, they discovered its anti-cancer properties quite by accident. Further research led them to isolate and characterize two of the most important chemicals in the plant as cancer-fighting medicines. The chemicals were given the names vincristine and vinblastine. Today, vinblastine has helped increase the chance of surviving childhood leukemia from 10% to 95%, while vincristine is used to treat Hodgkin's disease. This is a cancer of the lymph tissue, which is found in the liver and in bone marrow and elsewhere and is expressed in the number of white blood cells.

Thomas Hodgkin described the symptoms in 1832 (see Glossary).

MODERN RESEARCH AND THERAPEUTIC VALUE

In the 1950s, the scientists Robert Noble and Charles Beer at the University of Western Ontario in Canada discovered that extracts of the Madagascar periwinkle destroyed white blood cells. However, they had begun research on this plant based on folklore reports of anti-diabetic activity from a surgeon, Dr C. D. Johnson, living in Jamaica. The results of their research on the discovery of novel anti-cancer activity found in the chemical extracts were presented in March 1958 in a research symposium at the New York Academy of Science. Their paper had been submitted at the last minute by invitation of the conference organizer and was last of the evening's presentations. The symposium ran late and the Canadians presented their findings at midnight! The audience had by then dwindled to just a few listeners—mostly members of a team of researchers from the Eli Lily drug company.

During this period, Eli Lily was testing and screening hundreds of plant extracts each year in search of biological activity which might lead to the development of a new drug. Natural products chemist, Dr. Gordon H. Svoboda (1922–1994) at Eli Lily, had added the Madagascar periwinkle to the list of research subjects based on reports of use of periwinkle products for the treatment of diabetes in the Philippines during the Second World War. Independently of the Canadians, an extract of the plant was submitted for assay and Svoboda learned in early 1958 that the extract exhibited very high potency in anti-cancer tests. So, at the time that the Canadians' paper was presented in the spring of 1958, neither the Canadian research group nor the Eli Lily researchers knew of each other's work on the same plant. However, both groups had observed that their respective plant extracts lowered white blood cell counts in laboratory animals whilst they were looking for anti-diabetic effects. Since leukemia involves a proliferation of white blood cells, both teams made observations leading to the deduction that an agent which reduced the number of white blood cells might have potential value in the treatment of leukemia. From an extract of the plant, another scientist, Charles Beer, isolated one specific chemical compound. He named the active compound vincristine, which was also reported in a scientific paper in 1958 at a cancer research symposium. Needless to say, the Eli Lily research team was extremely interested in this Canadian researcher's work on the Madagascar periwinkle.

After an initial meeting, agreement on collaboration between the Eli Lily Company and the Canadian researchers was secured.

Thus, the race was on!

In March 1961, vincristine was approved by the United States Food and Drug Administration (FDA) as a chemotherapeutic agent for the treatment of Hodgkin's disease. Even more importantly, a second chemical compound, vinblastine, was isolated by Dr. Svoboda from the Madagascar periwinkle. In July 1963, this drug was approved in the United States for the treatment of childhood leukemia.

Within two years of the discovery of the compounds and their anti-tumor activity and approval as new drugs, Eli Lily had to secure and develop a significant supply of Madagascar periwinkle in order to make production commercially viable. This meant rapid development of farming operations. The extraction of just one gram of vinblastine from the Madagascar periwinkle required 2,000 lbs. of dried leaves. It should be noted that the chemical structures of these compounds are complex. Laboratory synthesis is tedious and expensive so natural supplies remain the best source.

Tens of thousands of cancer patients, especially those suffering with leukemia and lymphomas, have benefited from the drugs derived from this remarkable medicinal plant, the Madagascar periwinkle, better known to many in tropical climates as a weed and to the American gardener as an easy-to-grow ornamental. Serendipity led to an educated guess, and the hunch turned out to be correct.

The Alkaloids, Vincristine and Vinblastine

Although extracts of the Madagascan periwinkle had been used as a folk remedy for centuries, scientific studies were only carried out in the 1950s. Research revealed that extracts from the plant contained up to seventy different alkaloids, many of which are active on metabolic systems in humans. Alkaloids are very common natural compounds produced by a large variety of organisms including bacteria, fungi, green plants and animals. More examples of alkaloids are to be found in the chapters on 'Morphine: A Double-Edged Sword', 'Coffee, Wake Up and Smell the Aroma' and 'Cocaine'.

Two alkaloids of particular interest isolated from the Madagascan periwinkle were vincristine and vinblastine. The alkaloids are very similar to one another as their chemical structures are very closely aligned (see **Figure 2.23**).

Vincristine has the empirical formula $C_{46}H_{56}N_4O_{10}$. Vincristine contains many functional groups familiar to the student chemist: carbonyl groups, hydroxyl groups, carboxylic groups, aromatic rings and an indole ring. Despite this, structurally and chemically, vincristine and other alkaloids are extremely complex compounds.

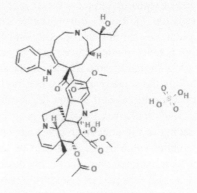

FIGURE 2.23 Vincristine (left) and vinblastine (right) in the form of the sulfate salt.

Source: PubMed

Vincristine and vinblastine are so similar that they differ by only one carbonyl group, present in an aldehyde functional group in the former, which is to be compared with a methyl group in the corresponding location in the latter. Can you spot the difference in Figure 2.23?

Furthermore, it should be noted that indole is an important building block for many naturally occurring alkaloids which include significant, complex chemical compounds extracted from the Madagascan periwinkle. Both indole and the indoles as a family of compounds are described in the chapter entitled 'Woad and Indigo', found in Part VII.

Isolation of Vincristine and Vinblastine from Plants

Fractional Distillation

In order to isolate vincristine and vinblastine from the source plant, much research had to be undertaken to find solvents that could dissolve these alkaloids. Thereafter, the first step in the process of extraction involves simply immersing the plant in the solvent. After filtration, the spent plant material is discarded and the solution retained. Then, the solution is concentrated by reducing the volume of solvent, which is achieved by fractional distillation. Distillation of the solution leads to the collection of a fraction that is finally refined by chromatography to yield a pure sample of vincristine.

Chromatography is treated in the chapter on 'An Asian Staple—Rice' while fractional distillation is described in the chapter entitled 'Central America's Humble Potato'. Incidentally, the process of steam distillation is described in the chapter 'European Lavender'.

Acid-Base Extraction

Alkaloids almost always contain at least one basic nitrogen atom so they can also be purified from crude extracts by acid-base extraction which is a process elaborated upon further in the chapters dealing with the de-caffeination of coffee and zwitterions, 'Coffee: Wake Up and Smell the Aroma' and 'Maca from the High Andes in South America'.

Questions

1. What property allows mixtures to be separated by fractional distillation?
   - density
   - boiling point
   - type of bonding

2. Explain hydrogen bonding in water and organic liquids and how it influences density and boiling point.

3. Whenever possible, why is melting point rather than boiling point used to check the purity of a sample of an organic substance?

4. From the structure of vinblastine, point out the indole sub units.

5. Explain simply in terms of functional groups the type of reactions you would expect vinblastine to undergo. Draw attention to any complications that may arise from the poly-functional nature of the molecule.

6. Give examples of other organic compounds which have the indole group as a building block and briefly describe their properties and commercial value.

7. Unlike most amines, indole is weakly basic as only very strong acids such as hydrochloric acid are able to protonate the nitrogen atom. Explain why indole differs from amines in this way.

8. Explain why the acid-base extraction is a useful technique in the purification of vinblastine and vincristine.

SUGGESTED FURTHER READING

M. Iwu. 2014. *Handbook of African Medicinal Plants*, 2nd Edition. CRC Press.

A. D. Osseo-Asare. 2014. *Bitter Roots. The Search for Healing Plants in Africa*, pp. 31–52. University of Chicago Press.

SAVING THE PACIFIC YEW TREE

Abstract: The amazing natural product, paclitaxel (commercially known as Taxol®), was isolated from a yew tree in the 1960s as an anticancer agent. The development of Taxol languished for years due to a lack of sustainable natural sources. Eventually, the full force of a joint effort by industry and government led to successful conservation of the Pacific Yew and to production of the drug through a process involving fermentation and semi-synthesis.

Organic Chemistry

- *isomers*
- *stereo-chemistry*
- *chirality*
- *nuclear magnetic resonance (nmr) spectroscopy.*

Context

- *terpenes in nature*
- *paclitaxel (Taxol).*

THE NATURE OF PACLITAXEL

Paclitaxel (Figure 2.24) is a terpenoid which was first isolated from the bark of Pacific Yew trees (*Taxus brevifolia*). Terpenes are hydrocarbons. Terpenes which contain additional functional groups are known as terpenoids (see also the chapter on 'A Plant from the East Indies, Camphor' in Part VI).

MEDICAL VALUE OF PACLITAXEL

Paclitaxel is a mitotic inhibitor used in cancer chemotherapy. It was approved by the Food and Drug Administration for treatment of drug-resistant ovarian and breast cancers and also is used in the treatment of lung cancer. Subsequently, it has become a major research tool of study in cancer therapy. Paclitaxel works by inhibiting mitosis: as the cells are unable to multiply, tumors are unable to grow.

MODERN-DAY PREPARATION OF PACLITAXEL

Today, the preparation of paclitaxel involves an elegant combination of measures. Large amounts of a precursor compound are isolated, which is followed by an additional semi-synthetic step to yield the final product. The source of the precursor is the English yew, *Taxus baccata*. The precursor, 10-deacetyl baccatin (10-DAB), is available from the needles of the tree in large quantity. The difference between 10-DAB and paclitaxel is simply that 10-DAB has no ester side-chain. The prepared side-chain is attached to the C-13 hydroxyl group of 10-DAB to obtain paclitaxel on a large scale. In this manner, paclitaxel is manufactured by semi-synthetic production from a natural precursor (Holton et al., 1994).

HISTORICAL PERSPECTIVE

The compound, Taxol®, was discovered in 1971 by Monroe and Wall who were seeking anticancer agents (Wani et al., 1971). Remarkably, many more years were to pass before further study at the National Cancer Institute (NCI) in the USA proceeded to clinical trials. The NCI showed great reluctance to pursue Taxol because its isolation was extremely difficult. The bark of the yew tree produced only small amounts of compound and, once stripped of its bark, the tree dies.

However, the program gained momentum due to efforts of Dr Matthew Suffness at NCI who found Taxol to be very active against melanoma. In 1978, it was also revealed that Taxol had the capability to cause considerable regression in mammary tumors.

In the early 1980s, NCI approached industry for support and Bristol Myers Squibb decided to develop Taxol into a clinically tested drug. New research showed Taxol could be obtained from the needles of the yew tree—ecologically much better than extraction from bark. Subsequently, Taxol was discovered as a fungal metabolite which provided the opportunity for large-scale production from fermentation.

TERPENES AND ISOPRENE AS BUILDING BLOCKS IN NATURE

Terpenes are a class of compounds that are common in nature. They are also described in the chapter on 'Cannabis and marijuana' in Part V and in the chapter on 'A Plant from the East Indies, Camphor' in Part VI.

Terpenes are present in tree resin from which turpentine can be extracted. Indeed, the very name, terpene, is derived from the word turpentine. Terpenes are major biochemical building blocks within nearly every living creature. Steroids, for example, are derivatives of the terpene known as squalene. Terpenes and terpenoids are also the primary constituents of the oils of many types of plants and flowers. These oils are used widely as natural flavor additives for food, as fragrances in perfumery and in both traditional and alternative medicines such as aromatherapy (see the chapters concerning 'Exotic Potions, Lotions and Oils' in Part VI).

Terpenes are polymeric compounds made up of a number of isoprene molecules, CH_2C (CH_3). $CH.CH_2$, which have a short five-carbon chain. Because all terpenes share a common building block, isoprene, terpenes can be categorized by how many isoprene units they include. The simplest molecule is a two-isoprene unit called a monoterpene, which has ten carbon atoms. Examples of derivatives of monoterpenes are camphor and menthol, which help clear mucus from sinuses when a person is suffering from a cold and pinene, which is used in wood varnish.

ISOMERS

Linkage between two isoprene molecules can occur in different ways involving either the 'head' or 'tail' of the molecule to form three distinct structural sequences that have the same empirical formula. As a consequence, each of these structurally different molecules has distinct chemistry. Each structurally different molecule is known as an isomer.

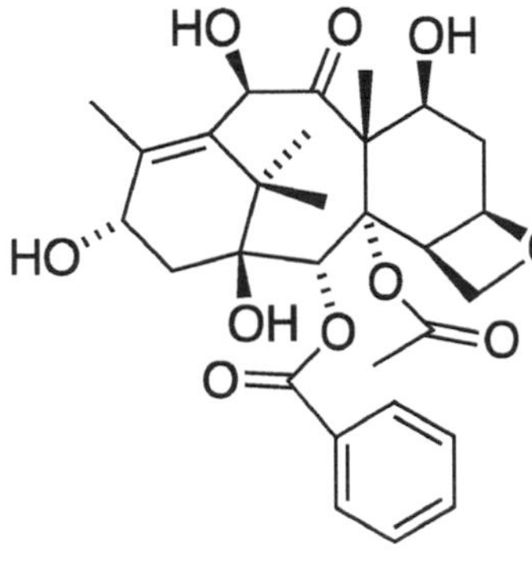

FIGURE 2.24 Paclitaxel.

Here is an isomer being formed from the head-to-head or 1–1 link

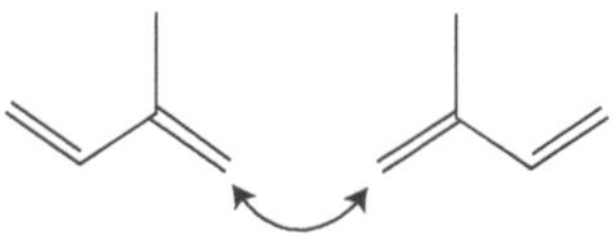

while this link is called a head-to-tail or 1–4 link.

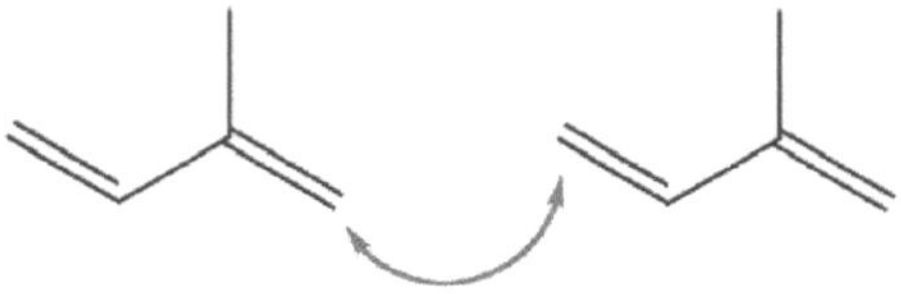

A rare linkage is called a tail-to-tail or 4–4 link.

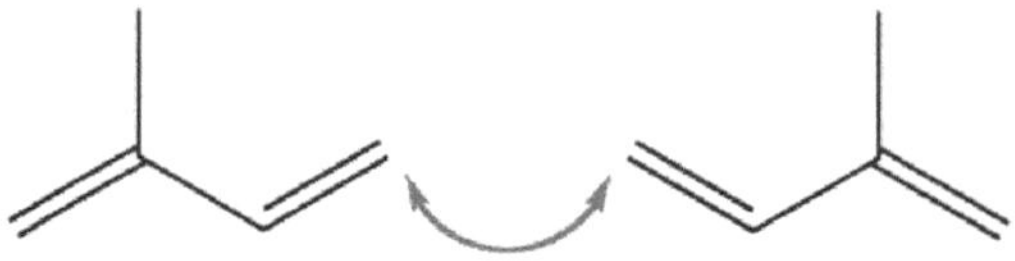

Isomers and Stereo-Chemistry

Physical rotation around the axis of a carbon–carbon double bond is not physically possible as the double bond is structurally rigid. This gives rise to stereo-chemistry because the relative orientation of the groups within a molecule can also give rise to different isomers.

The stereochemistry of carbon-carbon double bonds is usually called *cis* and *trans* isomerism but is also known as geometric isomerism (Figure 2.25). The terms *cis* means 'on the same side' and *trans* means 'across'.

An alternative, the E-Z notation, makes it possible to deal with more complex cases. Look at the atoms or groups attached to each of the carbon atoms in the double bond in the following example for E- and Z-butene (Figure 2.26). When the two methyl groups are on the same side of the C=C,

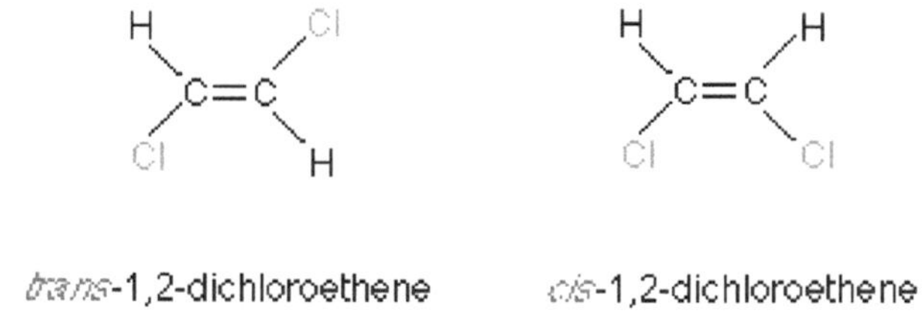
H
Cl
C=C
Cl
H
trans-1,2-dichloroethene
H
H
C=C
Cl
Cl
cis-1,2-dichloroethene

FIGURE 2.25 An example of two isomers: *cis* (same) and *trans* (opposite).

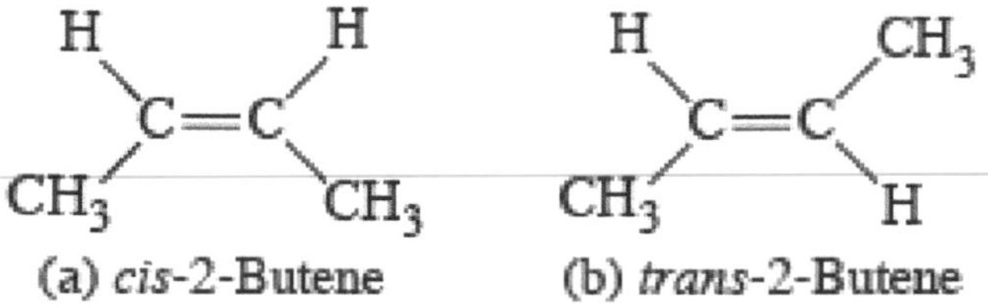
H
H
C=C
CH3
CH3
(a) cis-2-Butene
H
CH3
C=C
CH3
H
(b) trans-2-Butene

FIGURE 2.26 Stereoisomerism in alkenes.

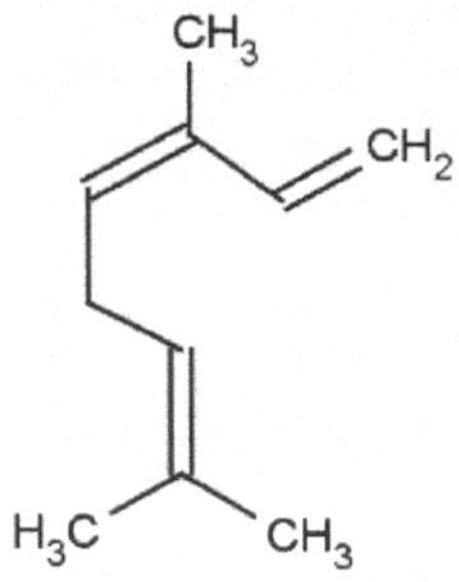

FIGURE 2.27   Structure of ocimene.

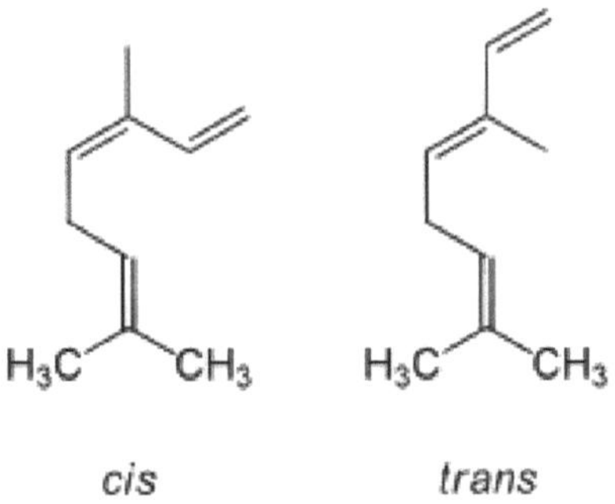

cis   *trans*

FIGURE 2.28   The geometric isomers of ocimene.

the isomer is described as Z, from the German word for together, *zusammen*. If not, it is E, from the German word for opposite, *entgegen*.

Ocimene (Figure 2.27) is another example of a compound that has geometric isomers found in the *cis* and *trans* configurations (Figure 2.28). It is extracted as an essential oil from the popular herb, basil (*Ocimum tenuiflorum*). Ocimene is classified as a linear terpene and possesses a pleasant odor and so finds application in perfumery.

ISOMERS AND CHIRALITY

The term chiral is used in general to describe an object that is not super-imposable on its mirror image. Human hands are an example. No matter how our two hands are orientated, it is impossible for all the major features of both hands to coincide in space. This difference in symmetry becomes obvious if a left-handed glove is placed on a right-handed glove.

The concept of chirality is extremely important in chemistry too. Each mirror image of a chiral molecule is a special type of isomer known as an enantiomer. A pair of enantiomers is often designated as 'right-handed' and 'left-handed'. Molecular chirality is of considerable interest in stereochemistry especially in organic chemistry and biochemistry where molecules can be large and complex. Spatial relationships between functional groups in different, large or complex molecules can fundamentally determine vital aspects of chemical interaction (see also the fascinating entry on the significance of the chirality of proteins and enzymes in the chapter on 'Wheat—Ancient and Modern' in Part II).

A simple example of a chiral molecule arises in a tetrahedral molecule in which all four substituents are different, e.g. fluoro chloro bromo methane, CHFClBr.

An enantiomer is a chiral molecule which also has the property of rotating the plane of polarized light. If the rotation of light is clockwise (as seen by a viewer toward whom the light is traveling), the enantiomer is labelled (+). Its mirror-image is labelled (−). This property helps scientists to study chirality and enantiomers.

NUCLEAR MAGNETIC RESONANCE (NMR) SPECTROSCOPY AND MOLECULAR STRUCTURE

The impact of nuclear magnetic resonance spectroscopy (abbreviated as NMR spectroscopy) on organic chemistry has been substantial. NMR spectroscopy is frequently used by chemists and biochemists to investigate the structure and hence the properties of organic molecules from small to large and complex such as proteins or nucleic acids or carbohydrates.

The technique relies on the phenomenon of nuclear magnetic resonance. Only nuclei having an odd number of protons or neutrons give a signal. Research often exploits the magnetic properties of the nuclei of the hydrogen atom ^1H, usually referred to as a proton in this context and an isotope of carbon, ^{13}C, which is about 1% abundant compared to the common isotope, ^{12}C.

A spinning charged atomic nucleus generates a tiny magnetic field. When an external magnetic field is applied, the difference between two energy levels (ΔE) is resolved arising from two possible directions of spin. The nuclei of some atoms align with the magnetic field while other nuclei line up against it.

Irradiation of the sample with energy in the radio-frequency band that corresponds to this small difference will cause excitation of those nuclei in the lower energy state (with the field) to the higher energy state. Resonant absorption of energy occurs at a frequency characteristic of the NMR-active atom, ^1H or the isotope of carbon, ^{13}C. The spinning nuclei that lie against the applied magnetic field will, of course, radiate energy at this frequency when they return to the ground or lower energy state in which the spinning nuclei are aligned with the field.

However, the magnetic field around a nucleus is also influenced slightly by the magnetic effects of the electrons present in other atoms within the molecule and most importantly by the distribution of the other atoms in the molecule. This of course is determined by the structure of the molecule. The energy difference between the spin states of the NMR active nucleus is therefore affected by the structure, which, in turn, alters the exact frequency of absorption of energy by a nucleus in a given externally applied magnetic field strength. Analysis of a spectrum of absorption signals from NMR active nuclei (such as a proton or the isotope of carbon, ^{13}C) can, in expert hands, reveal details of molecular structure including isomerism (see the chapter on 'Wheat—Ancient and Modern' in Part II, for the application of NMR spectroscopy to the understanding of the structure of proteins and enzymes). As there are fewer sharp absorption peaks in ^{13}C NMR spectra, they are usually more straightforward to interpret than proton spectra. Each functional group in organic chemistry, such as C-C, C=C, C-O, C=O, C-Cl, C-N, CNO, CHO, COOH or aromatic carbon rings, can be identified in ^{13}C NMR spectra from their distinctive chemical shift.

Fine details of the location of hydrogen atoms in a molecule can also be obtained from proton NMR spectroscopy as shown in the following spectrum of ethanol.

The chemical shift of the ^1H nuclei is the difference between its resonant frequency of absorption of energy and the frequency of absorption of proton nuclei in a standard such as tertramethyl silane which is set at zero on the NMR scale. The chemical shift of each group of protons, whether those present in CH_3, CH_2 or OH in this instance, is influenced by the charged electrons in their immediate proximity, in other words by the electrons in the functional group of which they are a part.

The signals from protons in the methyl and methylene groups are in turn split into multiple peaks or multiplets due to coupling with the spin of hydrogen atoms on adjacent carbon atoms. Spin-spin coupling is used to interpret NMR proton spectra through what is simply called the $n + 1$ rule, where

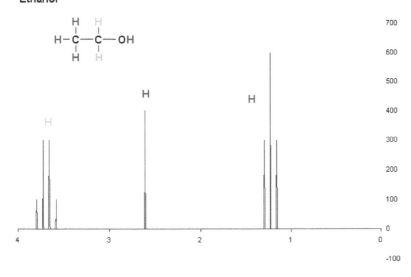

FIGURE 2.29 ¹H NMR spectrum of ethanol.

n is the number of hydrogen atoms on the adjacent carbon atoms. As an illustration in Figure 2.29 we show the ¹H NMR spectrum of ethanol.

When a peak in a proton NMR spectrum is split into two multiples, then there is one hydrogen atom on the neighboring carbon atoms.

When a peak in a proton NMR spectrum is split into three multiples, then there are two hydrogen atoms on the neighboring carbon atoms.

When a peak in a proton NMR spectrum is split into four multiples, then there are three hydrogen atoms on the neighboring carbon atoms.

Questions

1. Although the structure of a molecule of paclitaxel appears complex, its characteristics will be influenced by the functional groups present. Identify those functional groups and their chemical properties.
2. How might an ester side-chain be added at C–13 and what complications would you anticipate?
3. Which of the following exhibit(s) geometric isomerism?
   - but–2–ene
   - but-2–yne
   - phthalic acid based on a benzene ring with formula $C_6H_4 (COOH)_2$
4. Compare and contrast the chemical and physical properties of two structural isomers, namely ethanol and dimethyl ether.
5. Explain why spin-spin coupling is not observed in ¹³C NMR spectra.
6. Applying the $n + 1$ rule to the multiplets present in the proton NMR spectrum of ethanol shown in Figure 2.24, identify the signals from distinct hydrogen atoms in the CH_3, CH_2 and OH functional groups.
7. Explain in each case why tetramethylsilane, $(CH_3)_4Si$, or deuterated chloroform, $CDCl_3$, may be used as a standard in proton NMR spectroscopy.
8. Give an account of the value of nuclear magnetic resonance to society in general given its application in approaches to health scanning undertaken in hospitals.

REFERENCES

Key Paper describing the chemical synthesis of taxol:

R. A. Holton, C. Somoza, H. B. Kim, et al. 1994. *First Total Synthesis of Taxol. Functionalization of the B ring.* J Am Chem Soc 116: 1597–1598.

A seminal paper describing the isolation and structure of the anticancer compound Taxol from the Pacific Yew:

M. C. Wani, H. L. Taylor, M. E. Wall, P. Coggon, and A. T. McPhail. 1971. *Plant Antitumor Agents. VI. The Isolation and Structure of Taxol, a Novel Antileukemic and Antitumor Agent from Taxus Brevifolia.* J Am Chem Soc 93: 2325–2327.

VACCINES, ADJUVANTS AND ORGANIC CHEMISTRY

Abstract: *How carbohydrates known as saponins, extracted from the bark of the Chilean soap bark tree, have made a remarkable contribution to the development of vaccines effective against the latest contemporary pandemic, COVID-19.*

Organic Chemistry

- *hydrocarbons*
- *carbohydrates*
- *proteins*
- *glycosides*
- *RNA and DNA*
- *structural isomerism*
- *profound effect of an adjuvant on the effectiveness of a vaccine.*

Context

- *saponins*
- *terpenes.*

How the Chemistry of Natural Products Enhances Vaccines

Over eons of time, pandemics have regularly swept the world, ravaging humanity and changing the course of history. Several notable and relatively recent pandemics are presented.

The Black Death (1346–1353) traveled from Asia to Europe, where estimates indicate it wiped out over half of the population. It was caused by the bacterium *Yersinia pestis* and was spread by fleas on infected rodents.

European diseases, including smallpox, sweeping through Central and South America, contributed to the conquest of the Inca and Aztec civilizations by Hernán Cortés (1519) and Francisco Pizarro (1532) respectively.

The Great Plague of London (1665–1666) spread rapidly during the hot summer months. Fleas from plague-infected rodents were one of the main causes of transmission. By the time the plague ended about 15% of the population of London had died.

An estimated 500 million people across the globe were victims of Spanish Flu (1918–1920). Infection and lethality were enhanced by the cramped conditions of soldiers and the poor nutrition that many people experienced during World War I.

AIDS has claimed an estimated 35 million lives since HIV, the virus responsible, was first identified. The virus swept around the world during the late 20th century.

In those countries where COVID-19 has been fully reported, as of June 2023, there have been 767,000,000 cases and over 7,000,000 deaths.

Vaccines

During the modern era, when a pandemic does break out, public attention is inevitably drawn toward progress on cutting-edge vaccines such as those being deployed against the SARS-Cov-2 virus.

A vaccine can reduce the risk of a person contracting a disease by encouraging the body's immune system to:

- recognise an introduced antigen* as foreign
- produce antibodies, which are proteins, to combat the invading organism.

*Antigens are typically proteins or carbohydrates derived from the pathogen (virus or bacterium), against which an immune response is desired.

Once vaccinated, an individual may remain protected against a disease for years or even a lifetime. A vaccine can even eradicate a disease if it is distributed widely enough, as in the case of smallpox. Forestalling disease is a much better approach to protection than treating a disease with drugs after it has become established.

However, a vaccine is often comprised of *two* components; an antigen and an adjuvant. While a lot of public interest is placed on antigens, we do need to give credit to their unsung partners, adjuvants. Without them, vaccines really would not be as effective or as long-lasting or as widely available.

ADJUVANTS

An adjuvant is a substance that improves the effectiveness of a vaccine. Generally, an adjuvant is injected into the bloodstream with the antigen to help the immune system generate the antibodies required to combat the antigen. The adjuvant enhances the response of the immune system to the presence of an antigen (a foreign cell or virus or substance). However, great care must be exercised to avoid over-stimulation of the immune response by an adjuvant since this may lead to serious consequences as the immune system, operating through white blood cells (T cells), may attack and damage the body's own tissues. It is absolutely necessary to research thoroughly the use of a specific antigen in tandem with a particular adjuvant in order to create an effective vaccine.

THE DEVELOPMENT OF ADJUVANTS

Gaston Ramon (Figure 2.31) worked at the Pasteur Institute in Paris in the 1920s and is credited with the discovery of the medical value of adjuvants.

He noticed in 1925 that horses vaccinated against diphtheria showed stronger immune response when inflammation occurred at the site of injection. Ramon then set out to test a range of common materials and foodstuffs for their ability to cause irritation and inflammation as vaccine additives.

FIGURE 2.31   Gaston Ramon.

Source: Wellcome Collection gallery (2018–03–29) Library reference: ICV No 27511, Photo number: V0027053. Creative Commons Attribution 4.0 International license

On the basis of the simple belief that substances safe to ingest were safe to inject, Ramon demonstrated rather remarkably that various substances, from breadcrumbs to starch to jelly-like agar and soap, improved antibody responses in vaccinated animals. Some modern adjuvants are still based on related substances but are manufactured using more controlled and regulated methods.

A similarly serendipitous discovery followed a year later when Alexander Glenny, a British immunologist at the Wellcome laboratory in London, used aluminum salts in an attempt to purify the diphtheria protein. He found by accident that this preparation also enhanced antibody response compared to previous diphtheria vaccines. Over the years since then, inorganic compounds of aluminum have become one of the most widely used adjuvants. They include aluminum hydroxide (commonly used as an antacid to relieve indigestion and heartburn), aluminum phosphate and potassium aluminum sulfate (often found in baking powder). These adjuvants are used as components in many licensed, protein-based vaccines; including those against diphtheria, tetanus, hepatitis, pneumococcal and meningococcal diseases. Although we still do not fully understand how aluminum salts work as adjuvants, they are trusted and extensively used. Indeed, Sinopharm, a modern Chinese vaccine against COVID-19, contains a sterilized coronavirus as the antigen and an aluminum salt as the adjuvant.

By 1925, it had also been noted that the addition of saponin enhanced antibody responses to diseases such as diphtheria and tetanus. Since then, a number of adjuvants have been identified and successfully applied to enhance the efficiency of new vaccines.

A VACCINE AGAINST COVID-19

The spherically shaped SARS-Cov-2 virus responsible for COVID-19 consists essentially of a coiled double-strand of ribonulcleic acid (RNA), which contains its genetic code, encapsulated in a thin layer of protein for physical protection (Figure 2.32). There are protruding 'spikes' of a different protein which act as the tool by which the virus engages with the host cells in the victim in order to replicate and cause disease.

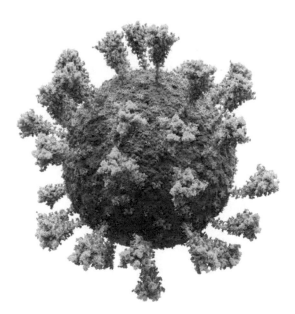

FIGURE 2.32   Model of the external structure of the Severe Acute Respiratory Syndrome CoronaVirus 2 (SARS-CoV-2).

Source: Credit: A. Solodovnikov, V Arkhipova. Creative Commons Attribution-Share Alike 4.0 International license

Adjuvants are used mainly in vaccines that depend for effect on the inclusion of sub units of protein that are present in the pathogen or its antigen. As molecular pathways underlying the medical effects of adjuvants have become a little clearer, it has been realized that the innate immune response is stimulated through recognition of the protein in such adjuvants. For more on proteins, see 'Wheat, ancient and modern'.

SAPONINS

Saponins are used as adjuvants because they are capable of triggering an intense immune response even in low doses. Saponins are a group of carbohydrates which are abundant in many plant species. They taste bitter, are often toxic and dissolve in both water and fats. When agitated in water, saponins produce a foamy lather, making them useful as soap. A variety of other commercial applications involve use of saponins in fire extinguishers and in carbonated beverages (hence the head on a mug of root beer).

Saponins belong to a class of substances known as triterpene glycosides.

Triterpenes

Terpenes are a class of hydrocarbon compounds which are common in nature (see 'Saving the Pacific Yew Tree'). Terpenes containing additional functional groups are known as terpenoids. Terpenes are major biochemical building blocks present in nearly every living creature. Terpenes are found in tree resin from which turpentine can be extracted. Indeed, the very name, terpene, is derived from the word, turpentine.

Terpenes are classified according to the empirical isoprene rule. Terpenes are polymeric compounds made up of a number of unsaturated isoprene molecules, $CH_2C(CH_3).CH.CH_2$, which have a short, five-carbon chain and two double bonds. The simplest terpene, known as a monoterpene, has ten carbon atoms and consists of two isoprene units. Examples of derivatives of monoterpenes are the monoterpenoids, camphor and menthol, which clear mucus from sinuses when a person is suffering from a cold.

Triterpenes, molecular formula $C_{30}H_{48}$, consist of three terpene segments, so there are six isoprene units. The triterpene, squalene, is presented in the following figure and is made up of six repeat units of isoprene. Steroids, when formed into carbon rings, are derivatives of the triterpene, squalene.

Triterpenes exist in a great variety of isomeric structures forming chains, branched chains and rings. In fact, five-ring (pentacyclic) structures are the most numerous and among them are steroids. Skeletons of carbon atoms making up six-ring (hexacyclic) triterpenes are particularly relevant to studies of saponins.

Glycoside

A glycoside is formed when a sugar molecule binds covalently to another group by the replacement of a hydroxyl group with a strong carbon–oxygen–carbon linkage, known as a glycoside bond. This covalent bond resembles the carbon–oxygen–carbon bond present in a molecule of ether (see 'Asian Staple: Rice').

Sucrose (table sugar) has a glycoside bond and is found universally in plants and in honey, which is derived by bees from plant material.

Lactose has a glycoside bond and is the sugar found in the animal kingdom and is present in small concentrations in blood plasma and in the milk of mammals.

Knowledge of the conformers of saponins, triterpene glycosides, will greatly help chemists to unravel the biochemistry behind the medical activity of these adjuvants. Remember, a conformer is a structural isomer which has a different spatial arrangement of atoms in the molecule.

NATURAL SOURCES OF SAPONIN

Certain plants, such as Chilean soapbark tree, are rich in saponins. Two species of the soapbark tree are endemic to South America, *Quillaja Saponaria* (Figure 2.33) and *Quillaja brasiliensis*. They were once abundant on the lower flanks of the Andes up to an altitude of 6,000 feet.

The inner bark of the Chilean soapbark tree, *Quillaja saponaria*, is a particularly good source of saponins. When pulverized and soaked in water at an extraction plant, the bark is transformed into a brown, bitter, frothy liquid that is the raw material of one of the most coveted vaccine adjuvants in the world, the saponin known informally as Quillaja Saponaria 21 or QS-21 for short since the formal systematic name of the compound is very long and unwieldy (Fleck et al., 2019).

Owing to potential risks to human health, only a few adjuvants have been approved by the US Food and Drug Administration. QS-21 is one of the newest (Zhu and Tuo, 2016). These unheralded helpers can turn a weak vaccine into an effective one or stretch a scarce vaccine supply during a pandemic. Not every vaccine requires an adjuvant, but many do. More than 200 vaccines are listed in the Milken Institute's COVID-19 vaccine tracker and approximately 40% of them are protein-based vaccines, which rarely work without an adjuvant.

Nine years ago, researchers estimated that global supply of *Quillaja* extract of pharmaceutical-grade was sufficient for just 6 million doses of vaccine. In this context, the salutary story of the Pacific yew tree is worth repeating, for its bark was the original source of the anti-cancer drug paclitaxel. The Pacific yew tree was threatened by large-scale harvesting in the 1980s (See 'Saving the Pacific Yew Tree'). Similarly, there is a risk of over exploitation of the *Quillaja* genus resulting in significant depletion in supplies of saponin. Naturally, pharmaceutical companies are concerned that the availability of vaccines could be threatened by failure to secure sufficient supplies of high-quality extract.

NEW BIOCHEMICAL APPROACHES

No matter how effective a vaccine against COVID-19 may be, it cannot put a dent in the pandemic unless it can be produced on a massive scale. However, the downside of dependence on an adjuvant

FIGURE 2.33   The leaves and flowers of *Quillaja Saponaria.*

Source: Franz Xaver, GFDL. GNU Free Documentation License, Creative Commons Attribution-Share Alike 3.0 Unported license

is that it adds one more link to a global supply chain of essential materials, one more crucial connection that can be broken; but there is some good cheer! An increase in the yield of adjuvant by a factor of a hundred has been achieved by resort to semi-synthetic methods. Cultured plant cells have been employed at Phyton Biotech to combat successfully over-exploitation of the Chilean soapbark tree for saponin.

Fully synthetic routes offer the most encouragement. Researchers have revealed the synthetic pathway that the Chilean soapbark tree, *Quillaja Saponaria*, follows to make saponins (Reed et al., 2023). Synthesis, immunological evaluation and analysis of the shape of conformers of new variants of saponin have also been reported (Ghirardello, 2019). These studies have not only provided expedient access to adjuvant-active saponins but have also yielded insights into activity at molecular level that correlate with in vivo medical properties.

MOLECULAR BIOCHEMISTRY

While the biological and medical effects of individual adjuvants are known, justifying widespread use in vaccines, details of the mechanisms in biochemistry at molecular level by which they influence the human immune system and relate to specific antigens are not well understood (Pulendran et al., 2021). Saponins, however, are known to be surfactants, which may shed some light on the particular mode of action of QS21 as an adjuvant (Liao, 2021).

SUMMARY

Through a multi-disciplinary approach involving chemists, biologists, medical practitioners and epidemiologists, scientists will continue to investigate the variety of complex mechanisms responsible for the action of adjuvants in influencing the human immune response. Resort to preparation of saponins by synthetic means will help the research community to better understand the mechanisms driving adjuvant activity and will permit the design of a completely new generation of adjuvants involving smaller molecules which will be more stable and easier to manipulate.

Growth in understanding of these processes will help in the development of new and safe vaccines for a wider range of afflictions but, no matter how effective a vaccine may be, it will not limit a pandemic unless it can be made (a) as effective as possible and (b) is used as economically as possible. It is reassuring that natural products continue to be both a source of scientific inspiration and a vital medicinal resource in the enhancement of a vaccine effective enough to tackle the COVID-19 pandemic.

REFERENCES

J. D. Fleck, A. H. Batti, F. P. da Silva, E. A. Troian, C. Olivaro, F. Ferreira, and S. Verza. 2019. *Saponins from Quillaja Saponaria and Quillaja Brasiliensis: Particular Chemical Characteristics and Biological Activities*. Molecules 24 (1): 171.

M. Ghirardello. 2019. *Exploiting Structure–Activity Relationships of QS-21 in the Design and Synthesis of Streamlined Saponin Vaccine Adjuvants*. Chem Commun 56: 719–722. DOI:10.1039/C9CC07781B

Y. Liao. 2021. *Saponin Surfactants Used in Drug Delivery Systems: A New Application for Natural Medicine Components*. Int J Pharm 603: 120709. DOI:10.1016/j.ijpharm.2021.120709

B. Pulendran, P.S. Arunachalam, and D. T. Higgins, 2021. *Emerging Concepts in the Science of Vaccine Adjuvants*. Nat Rev Drug Discov 20: 454–475.

J. Reed, A. Orme, A. El-Demerdash, C. Owen, and L. B. B. Martin. 2023. *Elucidation of the Pathway for Biosynthesis of Saponin Adjuvants from the Soapbark Tree*. Science 2023 March 24th; 379 (6638): 1252–1264. DOI:10.1126/Science.adf3727

D. Zhu and W. Tuo. 2016. *QS-21: A Potent Vaccine Adjuvant*. Nat Prod Chem Res 3 (4): 113.

Part III

Nutrition—Ancient and Modern Miracles

INTRODUCTION

Over millennia, disparate human populations have learned through an empirical approach to select plants that tasted good, offered nutritional value, were useful in cooking, could be ground up as a spice to improve flavor or could be utilized as a preservative.

In Part III, we mainly concentrate on the organic chemistry of cereals that have sustained civilizations across the globe. Historically, grain agriculture has been successful because grains can be stored, measured and transported more readily than many other kinds of food crops such as fresh fruits, roots and tubers. An excess of food could be produced and stored easily, which contributed to the creation of the first permanent settlements. Also, early civilizations developed techniques in planting and in planned farming in order to maximize yield, which has influenced modern, efficient practice in agronomy.

A summary of the chemistry in Part III follows.

Chapter	Organic Chemistry	Context
Chia and Quinoa	Alkanes, alkenes	Lipids
	Saturated and unsaturated	Gluten sensitivity
	Fatty acids	Human health issues
	Trans-esterification	Vegetable oil and Biodiesel
Wheat	Amino-acids, peptide link	NMR spectroscopy
	Condensation reaction	Artificial intelligence + fast computers
	Fermentation	Shikimic acid pathway
	Hydrolysis	Phenylpropenoids
	Proteins and Enzymes	Crop yield and Green Issues

(Continued)

DOI: 10.1201/9781032664927-3

(Continued)

Chapter	Organic Chemistry	Context
Rice	Aldehydes (alkanals)	Oxygen in the organic ring
	Ketones (alkanones)	Mono, di and polysaccharides
	Chromatography—	Anthraquinones, glycosides
	Paper, Thin Layer, Gas	Optical isomerism and pharmaceuticals
	Ion exchange	
	Carbohydrates	
	Optical isomerism	
Cordyceps	Fermentation	Role of NAD+ and NADH
	Sugars to alcohol and CO_2	in redox reactions
	Redox reactions	

REDISCOVERING TRADITIONAL GRAINS OF THE AMERICAS—CHIA AND QUINOA

Abstract: *Chia from Central America and quinoa from South America lead a modern renaissance in appreciation of the value of ancient seed grains. These grains have high levels of desirable fatty acids, which are usually found only in fish. They are not cereal crops and so the grains are free of gluten.*

Organic Chemistry

- *alkanes and linear carbon polymers*
- *alkenes*
- *carboxylic acids*
- *saturated and unsaturated fatty acids*
- *cis and trans isomerism*
- *human health issues*
- *vegetable oil and biodiesel.*

Context

- *lipids also known as fatty acids.*

CHIA

Chia, *Salvia hispanica*, is a flowering herb. The plant belongs to the mint family and is native to Central America. Chia yields a grain-like seed, which provides a rich vegetable source of omega-3 fatty acids. Chia has been domesticated for the first time in modern history to supply economically viable quantities for world demand. Chia, literally meaning 'oily', originated in Mexico and was cultivated by the Aztecs. It is grown commercially today in regions of Central and South America. The useful parts of chia are the seed and the sprout. The seed contains an important lipid, an omega-3 fatty acid known as α-linolenic acid (see Figure 3.1), together with significant concentrations of dietary fiber, protein, calcium, magnesium, iron and antioxidants.

Scientific reports lay emphasis upon the importance of omega-3 fatty acids for our general health. It has been postulated that a deficiency in omega-3 fatty acids in the diet could lead to a greater risk of heart disease, cancer and cognitive defects. The richest dietary source of omega-3 fatty acids is found in fish. However, eating large amounts of fish has been linked with unwanted intakes of heavy metal pollutants—mainly mercury—and possibly other organic pollutants, such as poly-chlorinated biphenyls. Chia, therefore, offers a reasonable plant alternative.

QUINOA

Quinoa is another ancient grain from the plant, *Chenopodium quinoa*, whose value is often overlooked. *Chenopodium quinoa* originated in the cooler Andean regions of Ecuador, Bolivia, Colombia and Peru. It was successfully domesticated 3,000 to 4,000 years ago for human consumption. The Incas, who held the crop to be sacred, referred to quinoa as *chisaya mama* or 'mother of all grains'. It was the Inca emperor who would traditionally sow the first seeds of the season using 'golden implements'.

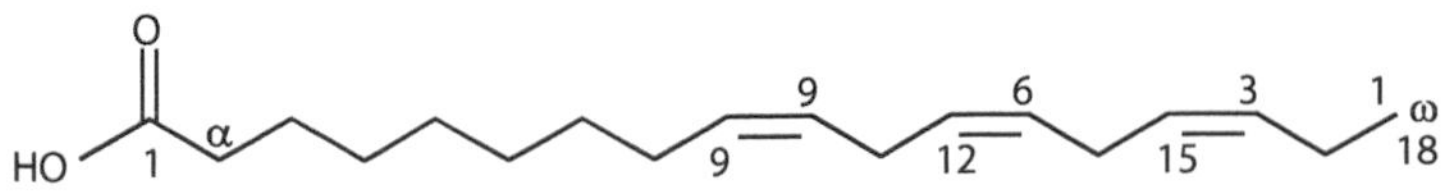

FIGURE 3.1 The structure of the omega fatty acid, α-linolenic acid.

During the European conquest of South America, the Spanish colonists scorned quinoa as 'food for Indians' and even actively suppressed its cultivation. The name is derived from the Spanish spelling of the Quechua word *kinwa*. Its nutrient value is very high compared with other modern cereals.

The grains contain essential amino acids. An *essential* amino acid or *indispensable* amino acid cannot be synthesized de novo (from scratch) by humans and therefore must be supplied by the diet. Amino acids are the building blocks for peptides and proteins (see the chapter on 'Wheat—ancient and modern' in Part II.

The grains also contain significant amounts of the minerals calcium, phosphorus and iron. After harvest, the grains need to be processed to remove the coating containing bitter-tasting saponins and are generally cooked in the same way that rice is prepared.

Quinoa is a grain-like crop grown primarily for its edible seeds. Although quinoa seeds resemble grain in appearance and in characteristics as a food, quinoa is not a member of the grass family and, therefore, is not considered as a true cereal. In fact, quinoa is closely related to species such as beets, spinach and tumbleweeds. Its nutrient value, however, compares well with that of common cereals.

Quinoa was of great nutritional importance in pre-Columbian Andean civilizations, second only to the potato, which was followed in importance by maize. In contemporary times, this crop has become highly appreciated for its nutritional value, as its protein content is very high (18%). Unlike wheat or rice (which are low in lysine), quinoa contains a balanced set of essential amino acids for humans, making it a complete protein source. Today, quinoa seeds may be ground into flour for baking bread or cakes. Quinoa seeds are often simmered gently in water, rather as rice grains are, to make a porridge-like breakfast dish or may be served in savory salads. Even more importantly, quinoa is gluten-free and considered easy to digest, offering an important nutrient for those suffering from gluten-related diseases.

GLUTEN SENSITIVITY

Gluten is a mixture of the proteins gliadin and glutenin, which contribute to the elasticity of dough. Wheat protein and the wheat starch are easily digested by nearly 99% of the human population. However, some research studies in several countries worldwide suggest that approximately 0.5–1% of these populations may have a genetic disorder that gives rise to celiac disease. This condition is caused by an adverse reaction of the immune system to gliadin present in gluten. Gliadin is a monomeric protein which is soluble in water whereas gluten is polymeric and insoluble in water. Over time, the digestive tract of the small intestine becomes inflamed and serious damage to the lining of the intestinal wall can result. In turn, the ability of the human body to absorb nutrients is affected with resultant discomfort and a general failure to thrive. Hence, susceptible people need a gluten-free diet.

As noted, quinoa is gluten-free and is being considered a possible crop in NASA's Controlled Ecological Life Support System for long-duration human occupied spaceflights.

LIPIDS

Lipids are a class of organic compounds, which, as fatty carboxylic acids, are insoluble in water yet soluble in many organic solvents. They include natural oils, waxes and steroids. Lipids, along with carbohydrates and proteins, are important components of all plant and animal cells.

Lipids of medium chain-length are often referred to by common names, which reflect their origin; for instance, palmitic acid, $CH_3(CH_2)_{14}COOH$, comes from the oil palm grown extensively in the East Indies, whereas stearic acid, $CH_3(CH_2)_{16}COOH$, is a component of soap.

Naturally occurring fatty acids may be saturated (and have no double bonds, like those described earlier) or unsaturated. Two examples of unsaturated lipids are presented in Figure 3.2. You can see that these molecules are made up from essentially the functional groups of a carboxylic acid, a long-chain alkane and an alkene. These two compounds are geometric isomers. They have the same

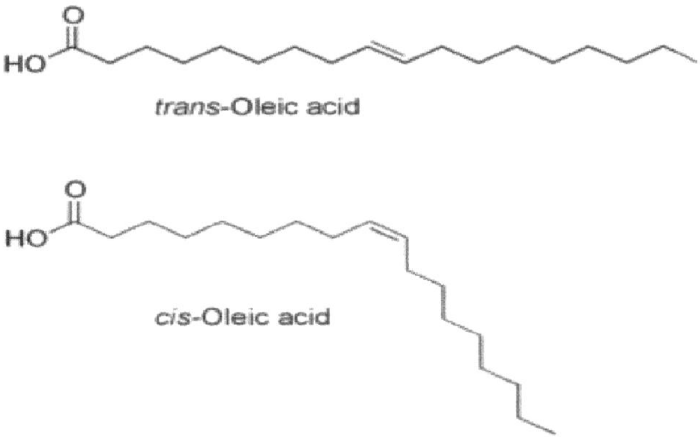

FIGURE 3.2 Examples of two unsaturated long chain fatty acids. The difference is in the substitution around the double bond: either *cis* or *trans*.

molecular weight and an identical molecular formula, but there is a difference in the orientation around the double bond known as *cis* and *trans*, as explained earlier (see Figure 2.25)

Esters of fatty acids are classed also as lipids. As a consequence of the size of the molecules and the presence of these well-understood functional groups, lipids exhibit a great deal of structural variety.

FATS, OILS AND HUMAN HEALTH

Fatty acids with three carboxyl groups form tri-esters with glycerol (propane-1, 2, 3-triol). Water molecules are formed along with the ester in what is known as a condensation reaction, described further in the next chapter. These triglyceryl esters (or triglycerides) compose the class of lipids known as fats and oils (for more about esters see the chapter on the foxglove entitled 'A Steroid in your Garden' in Part II). Triglyceryl esters are found in both plants and animals and form one of the major items in the human diet. Triglycerides that are solid at room temperature are classified as fats and occur predominantly in animals. Those triglycerides that are liquid are called oils and are found chiefly in plants, although triglycerides from fish are largely oils too. As might be expected, fats have a predominance of saturated fatty acid in the molecule whereas oils are composed largely of unsaturated acids.

Fats effectively hold a lot of energy in the extensive covalent bonds of the alkane chain, which is why mammals store fats in their tissues. When the body needs energy, fats can be oxidized to release it. Excess sugar in the diet is converted to fat. Low fat and low sugar regimes, coupled with food based on plant products, are of value in a healthy diet, particularly in developed countries of the world, where physical exercise among inhabitants is at a premium.

The animal fat in human food largely consists of saturated triglycerides, but it also contains a small amount of cholesterol. In contrast, cholesterol is not found in significant quantity in plant sources, which has ramifications for the human diet, since the ingestion of animal fat is one factor that influences blood cholesterol levels. Cholesterol is produced partly in the human body and is partly absorbed from the animal fats found in eggs, dairy products and meat.

As a complex lipid, cholesterol is only slightly soluble in water and, therefore, it dissolves in the bloodstream only at exceedingly small concentrations and can easily be brought out of solution. An accumulation of material (sometimes called plaque), including cholesterol and fatty acids, can build up on the inner wall of an artery and may narrow the artery, thereby restricting blood flow. High blood pressure, greater risk of heart attack and increased susceptibility to heart disease can result.

Plant (vegetable) oils are extensively converted on an industrial scale to solid triglycerides by partial hydrogenation of their unsaturated components. The hydrogenation of vegetable oils to produce semisolid products, which are used in the production of processed foods such as margarine or ice cream, has had unintended consequences. Although hydrogenation imparts desirable features in processed food, when compared with those of naturally occurring liquid vegetable oils, such as spreadability, texture and lengthened shelf life, it introduces some serious health problems. These occur when the *cis* double bonds in the fatty acid chains are not completely saturated in the hydrogenation process. The catalysts used to effect the addition of hydrogen cause isomerization of the remaining double bonds from the *cis* to the *trans* configuration. Unnatural fats in the *trans* configuration appear to be associated with an increased incidence of heart disease, cancer, diabetes and obesity, as well as difficulties in immune response. Because of these concerns, attention has turned to the process of trans-esterification. Natural esters in vegetable oils are converted into different esters with much higher molecular weight which melt at higher temperatures. These compounds provide the benefit of a smooth consistency to margarine yet avoid the *trans* isomers associated with health issues.

Trans-esterification simply involves reacting an ester with an alcohol to obtain a new ester, that is, the alcohol stem of the original ester is replaced with a different alcohol stem as in the following example.

$$CH_3COOCH_2CH_3 + CH_3CH_2CH_2OH = CH_3COOCH_2CH_2CH_3 + CH_3CH_2OH$$

Ethyl ethanoate combines with propanol in a reaction which is reversible to produce propyl ethanoate and ethanol. More generally, this reaction can be represented by

$$CH_3COOR + {}^1ROH = CH_3\,COO^1R + ROH$$

where R and ^{1}R are different stems of alcohols.

WAXES

Waxes are lipids and lipids are esters of fatty acids with long-chain monohydric alcohols (one hydroxyl group). Natural waxes are often mixtures of such esters and may also contain hydrocarbons.

Waxes are widely distributed in nature. The leaves and fruits of many plants have waxy coatings, which may protect them from dehydration and small predators. The feathers of birds and the fur of some animals have similar coatings, which serve as a water repellent.

The chemical formulae of two well-known waxes, beeswax and carnuba wax, are respectively $CH_3\,(CH_2)_{24}CO_2-(CH_2)_{29}CH_3$ and $CH_3\,(CH_2)_{30}CO_2-(CH_2)_{33}CH_3$. Beeswax has many uses, which include the manufacture of fine candles and is also applied to decorate and protect fine furniture. Carnuba wax is valued for its toughness and water resistance as a polish and as a sealant.

VEGETABLE OIL AND BIODIESEL

Biodiesel is a renewable fuel used in diesel engines. Biodiesel is made from vegetable oil or animal fat. A particular advantage of biodiesel is that it can be produced by recycling vegetable oils of poor quality or processed from oils which are left over from human activity, such as cooking oil. Animal fat from food processing plants can be used also.

Essentially, biodiesel is a mixture of the methyl and ethyl esters of fatty acids, which are produced by reacting methanol or ethanol with the triglycerides of oils and fats. This reaction is another example of trans-esterification.

Usually, biodiesel is not used in pure form but is mixed with conventional diesel, obtained from the cracking of petroleum. As an example, fuel denoted as B20 would contain 20% biodiesel. It should be noted that vegetable oil, as a renewable product, could be obtained from crops, such as the oil seed rape, grown on a much greater scale than at present, in order to provide a means of reducing human dependency on petroleum, which is not replaceable.

Questions

1. The richest dietary source of omega-3 fatty acids is fish. However, eating large amounts of fish has been linked with unwanted intakes of heavy metal pollutants such as mercury (and possibly other pollutants, such as poly-chlorinated biphenyls). How could how trace concentrations of heavy metal pollutants in the open ocean be intensified to significant levels for human health through the natural food chain of fish?
2. Explain why lipids are hydrophobic.
3. In organic chemistry, the determination of melting point is a well-known technique for establishing the purity of a sample dependent on the fact that the melting points of many organic compounds are so precise. Lipids which are saturated acids have higher melting points than lipids of corresponding size which are unsaturated acids. Explain why this is so.
4. Given the functional groups present in saturated lipids and unsaturated lipids, provide different examples of both and describe the chemical properties that you would expect them to possess.
5. Explain why trans fatty acids are considered harmful to human health.
6. Give a balanced account of how reasonable it would be to produce biodiesel on a very large scale from vegetable oils bearing in mind alternative options in agricultural practice. Make sure you also address the greenhouse effect and carbon neutrality in your answer.
7. Figure 3.1 presents an example of an unsaturated fatty acid. What chemistry would you suggest to create the respective saturated molecule? Give the chemical equation.
8. What is the chemical difference between an unsaturated and a saturated fatty acid? What are the implications for human health?

SUGGESTED FURTHER READING

R. Ayerza Jr. and W. Coates. 2005. *Chia*. University of Arizona Press.
J. F. Scheer. 2001. *The Magic of Chia. Revival of an Ancient Wonder Food*. Frog Ltd.

WHEAT—ANCIENT AND MODERN

Abstract: At the dawn of civilization in the Fertile Crescent of the Middle East, development of ancient strains of wheat led to the spread of its cultivation throughout the world. Cereal production is related to major contemporary questions concerning human population growth, genetic modification of crops and use of fertilizers.

Organic Chemistry

- *the condensation reaction and hydrolysis*
- *amino acids and the peptide link as a building block*
- *proteins*
- *enzymes*
- *the molecular structure of proteins*
- *the roles of NMR spectroscopy and artificial intelligence in structure determination*
- *chiral molecules and the organic chemistry of the origin of life.*

Context

- *sustainable practice: society's challenges in enhancing agricultural production—crop yield and green issues including the genetic modification of crops*
- *DNA and RNA*
- *shikimic acid pathway in plant metabolism*
- *phenylpropanoids.*

POSSIBLE ORIGINS OF ANCIENT WHEAT

The word *cereal* derives from *Ceres*, the name of the Roman goddess of harvest and agriculture. Cereals are grasses and are also known as *Gramineae*. They are cultivated for the edible components of their grain composed of the endosperm, germ and bran.

Archaeological findings indicate that ancient wheat varieties first occurred in parts of Turkey, Lebanon, Syria, Israel, Egypt and Ethiopia. Cultivation of wheat began to spread beyond the Fertile Crescent, between the major rivers of Syria and Iraq, after about 8,000 BC (Lev-Yadun et al., 2000). Jared Diamond, in his excellent book, *Guns, Germs and Steel*, traces the spread of cultivated wheat from the Fertile Crescent in about 8,500 BC; reaching Greece, Cyprus and India by 6,500 BC; Egypt shortly thereafter; followed by introduction in Germany and Spain around 5,000 BC. By 3,000 BC, wheat had reached England and Scandinavia. A millennium later it appeared in China.

MODERN WHEAT VARIETIES

Common wheat (*Triticum aestivum*), typically used in bread, is the most widely cultivated species in the world. Durum wheat (*Triticum durum*) is the second most widely cultivated wheat. Einkorn (*Triticum monococcum*) was domesticated at the same time as emmer wheat (*Triticum dicoccum*) but neither of these species is in widespread use. Spelt (*Triticum spelta*) is cultivated in limited quantities. Wheat is widely cultivated as a cash crop because it produces a good yield per unit area, grows well in a temperate climate and has a moderately short growing season of about 120 days.

NUTRITIONAL IMPORTANCE AND THE GRINDING OF WHEAT

Wheat whole grain is a concentrated source of vitamins, minerals and protein. However, when whole grain is 'refined' by grinding to remove the fibrous parts of bran and germ, the endosperm remaining consists essentially only of starch, a carbohydrate.

Gluten present in wheat grain is strong and elastic and enables the dough made from it to trap carbon dioxide during the leavening of bread. Gluten sensitivity is discussed in the chapter on 'Rediscovering Traditional Grains of the Americas—Chia and Quinoa'.

IMPORTANCE OF PROTEINS

Protein is found in the whole grains of a variety of cereal plants: wheat, oats, rye, millet, maize (corn) and rice. Proteins are essential nutrients for the human body and can also serve as an energy source. Protein can be found in all cells of the body. It is the major structural component of cells, especially those of muscle.

Proteins are polymer chains made from amino acids linked together by peptide bonds. During human digestion, proteins are broken down into smaller polypeptide chains. This process of release of essential amino acids is crucial since they cannot be biosynthesized by the body.

Enzymes are proteins too, but they are of a particular type with a highly specialized function. An enzyme is a metabolic catalyst that will only act with a given substrate, which has to fit exactly into an active chemical site on the enzyme. This is rather like a skillfully cut key fitting precisely into a well-made lock such that the lock can be operated at will (Figure 3.3). If the substrate fits exactly into the active site in each one of three dimensions in space, metabolism will proceed; otherwise it cannot.

The substrate is held in place on the active site without chemical binding by weak close-range attraction, due to a combination of hydrogen bonding and Van Der Waals forces. Once the substrate has been captured in this way, the substrate can react more readily with other compounds present. This is because the binding of the substrate to the enzyme causes redistribution of the electrons in the chemical bonds of the substrate. This event triggers reactions in the substrate, leading to the formation of products, which are released, thus allowing the active site of the enzyme to engage in another reaction cycle.

PEPTIDE BOND

A peptide bond (Figure 3.4) is a covalent chemical bond formed between two molecules when the carboxyl group of one molecule reacts with the amino group of another molecule, causing the

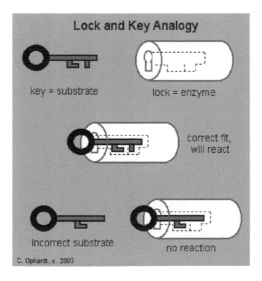

FIGURE 3.3 Schematic illustration of an active site in an enzyme—the lock and key model.

Source: released into the public domain worldwide by its author, Waikwanlai

$$2NH_2COOH = NH_2CONHCOOH + H_2O$$

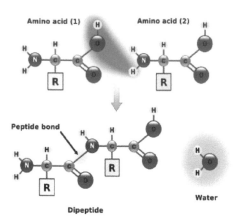

FIGURE 3.4 Formation of the peptide bond and formation of a dipeptide.

Source: released into the public domain by author

release of a molecule of water. This reaction occurs because amino acids are amphoteric; that is, they possess both alkaline properties (due to an amine functional group) and acidic properties (due to a carboxylic functional group) at one and the same time (see also the chapters on 'Tobacco: a profound impact on the World' for more information on amines, and 'Europe Solves a Headache' for more material on carboxylic acids).

The following reaction between two molecules to form the most simple of amino acid illustrates.

$$2NH_2COOH = NH_2CONHCOOH + H_2O$$

Proteins are only built from alpha (α) amino acids in which the functional groups, amine and carboxylic acid, are covalently bonded to the *same* carbon atom (identified as the α carbon).

The resulting group, C(O)NH, is called a peptide bond or link. Peptide bonds can be formed repeatedly between the carboxyl functional group and the amino functional group of neighboring amino acids. Proteins are polymers formed from chains of peptide bonds.

THE CONDENSATION REACTION

The process of forming a peptide bond is an example of a condensation reaction. In a condensation reaction, the molecules of two compounds combine together with the elimination of a molecule of water. There are many examples of condensation reactions in organic chemistry beyond the formation of polymers such as proteins. For instance, ethanoic or acetic acid forms acetic anhydride and water by the condensation reaction

$$2CH_3COOH = (CH_3CO)_2O + H_2O.$$

This reaction in reverse would be an example of hydrolysis.

PROTEINS AND AMINO ACIDS

As we have seen earlier in this chapter, proteins are polymers made up from building blocks known as amino acids that are linked together by peptide bonds. Over 500 different amino acid structures have been identified in nature. However, the human genetic code relates to only twenty of these. In other words, every protein in the human body consists of some combination of only these twenty amino acids. They can be sorted into two types: essential and non-essential. Essential amino acids must be obtained from the diet, whereas non-essential amino acids are those which the human body is capable of synthesizing.

The names of *essential amino acids* are histidine, isoleucine, leucine, lysine (Figure 3.5), methionine (Figure 3.20), phenylalanine, threonine, tryptophan, and valine.

The *non-essential amino acids* are as follows: alanine, arginine, asparagine, aspartate, cysteine (Figure 3.21), glutamic acid, glutamine, glycine, proline, serine (Figure 3.6) and tyrosine (Figure 3.11).

Amino acids found in the body cannot be stored in the same manner as fat or starch and must be obtained regularly from foods which contain proteins. Protein intake is therefore vital. The amino acids are obtained from the breakdown of proteins, which are present in the tissues of either animals or plants. The legume family is an example of a good source of dietary protein from plants.

AMINO ACIDS, PROTEINS AND DNA

The genetic code of an organism is contained in the unique sequence of peptides in the polypeptide, deoxyribonucleic acid (DNA). Whilst over 500 different amino acids have been identified in nature, the human genetic code relates to only twenty of these. In other words, every protein in the human body consists of some combination of only these twenty amino acids. In fact, every one of the proteins found in plants and animals is made up from sequences involving only twenty amino acids. The exact sequence of amino acids in a protein is copied, as from a template, from a specific sequence of nucleotides located along a part of the huge chain in DNA.

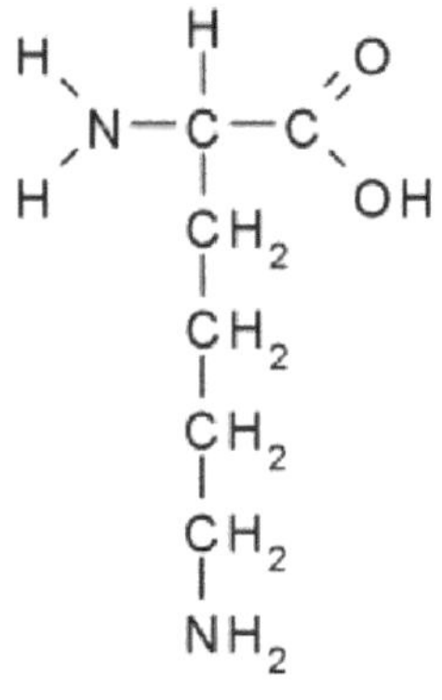

FIGURE 3.5 Molecular structure of lysine—an example of an essential amino acid.

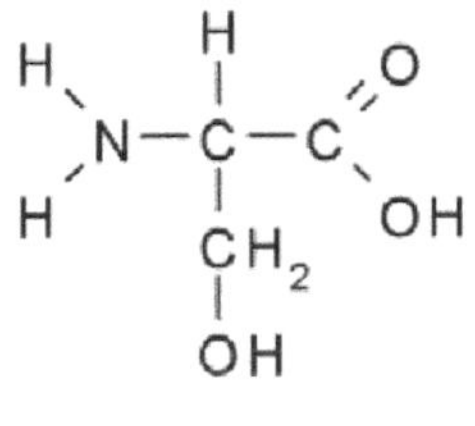

FIGURE 3.6 The molecular structure of serine—an example of a non-essential amino acid.

Elucidating the structure of the proteins and DNA of different organisms, including those of plants, has become a focus of scientific endeavor over recent years. Two of the techniques used in the quest are *nuclear magnetic resonance spectroscopy (NMR)* and, latterly, *artificial intelligence*

An important goal of research is to obtain three-dimensional structures of different proteins in high resolution. This can be achieved by the use of x-ray crystallography or by another important technique, nuclear magnetic resonance spectroscopy (NMR; see the chapter on the 'Saving the Pacific Yew' for an explanation of the principles behind NMR spectroscopy).

The active site of an enzyme, which is a protein, can be better understood using NMR. For example, let us consider the way modern drugs interact with our body. Many effective drugs are enzyme inhibitors (see Glossary). The mechanism of action of these drugs can be described using the model of a lock and key. If the details of the structure of the protein are examined using NMR and are compared before and after interaction with the trial drug, then the nature and site of biochemical activity within the molecule can be deduced. NMR spectroscopy has also been used to help to elucidate the structures and isomers of other complex molecules, such as nucleic acids and carbohydrates.

The very large scale of protein polymers and the huge number of proteins possible from just twenty key amino acids means that identification of the structure and shape of even a small number of proteins has been extremely challenging. In fact, since the determination of the structure of hemoglobin in 1962, relatively few have been solved. In the modern era, *artificial intelligence* and *fast computers* are of profound value since it is now possible to predict the shape of a complex protein in minutes from knowledge of the sequence of basic building blocks of amino acids present (DeepMind Technologies, 2020). Easier identification of the shape and function of proteins within the human body is bound to sharply accelerate medical advancement in the development of many new and effective therapeutic drugs.

FASCINATING INFLUENCE OF CHIRALITY ON LIFE SYSTEMS ON EARTH

In the chapter on 'Saving the Pacific Yew', attention was given to the concept of chirality in molecules, as this is of fundamental importance in natural products chemistry.

It is fascinating to note that only left-handed amino acids are found in the life systems of plants and animals. As a consequence, only left-handed proteins (and enzymes) are found in the natural world. Outside the natural world, nucleotides are very uncommon, but when they are found they exist in both left-handed and right-handed molecular structures in approximately a ratio of 1:1.

Life on Earth could have emerged from either left-handed or right-handed amino acids but not both, as metabolic processes and pathways are structure specific and are not possible with a mix of left- and right-handed proteins and enzymes.

Is this chance or design? No one knows for certain! It is a conundrum, which has intrigued scientists for many years and has stimulated much debate. Although unproven theories abound, some perhaps rather close to speculation, sheer chance seems the most likely explanation. A suggestion for further reading is given at the end of the chapter concerning the 'Alanine World' hypothesis (proposed by Kubyshkin et al., 2019). This hypothesis explains the evolutionary choice of amino acids from a chemical point of view. The selection of building blocks (amino acids) for protein synthesis is limited to those derivatives of alanine that are suitable for building the dominant secondary structures in the chemistry of life such as the α-helix of the genetic code. It is intriguing that most of the twenty amino acids (essential and non-essential) may be regarded as chemical derivatives of alanine.

THE SHIKIMIC ACID PATHWAY IN PHYTO-CHEMISTRY—A ROUTE TO AROMATIC AMINO ACIDS

The shikimic acid pathway is encountered in the plant kingdom; in green plants, in bacteria and in fungi but *not* in the animal kingdom. It deserves mention because the shikimic acid pathway is a significant feature in phyto-chemistry as a route to aromatic amino acids—such as phenylalanine (Figure 3.8)—and proteins.

The molecular structure of shikimic acid is presented in (Figure 3.9).

The shikimic acid pathway (Figure 3.10) involves seven enzymes acting in sequential steps to catalyze the biosynthesis of three aromatic amino acids; phenylalanine, tryptophan or tyrosine. The structures of these key essential aromatic amino acids are shown in Figure 3.11.

In the schematic illustration, the shikimic pathway first generates an intermediate compound, a phenylpropanoid called tyrosine, which in turn may be converted to amino acids, peptides and proteins.

Phenylpropanoids (Figure 3.7) are naturally occurring phenolic compounds (Figure 3.8).

Phenylpropanoids are essential building blocks that are converted by plants into

- amino acids, proteins and various secondary metabolites (see Glossary)
- pigments
- lignin for physical strength, for ultraviolet light protection and for defense against herbivores and insect attack.

Phenylpropanoids are found throughout the plant kingdom. They serve as essential components in a number of structural polymers, provide protection from ultraviolet light, defend against herbivores and pathogens and promote plant-pollinator interactions as floral pigments and scents.

Please refer to the chapters, 'Colorful Chemistry: A Natural Palette of Plant Dyes and Pigments', for more on pigments and to 'The Wonderful World of Wood' for more on lignin.

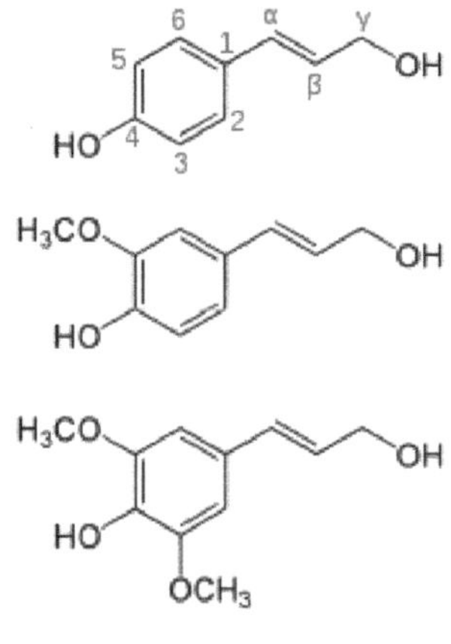

FIGURE 3.7 Three important phenylpropanoid compounds: paracoumaryl alcohol (1), coniferyl alcohol (2) and sinapyl alcohol (3).

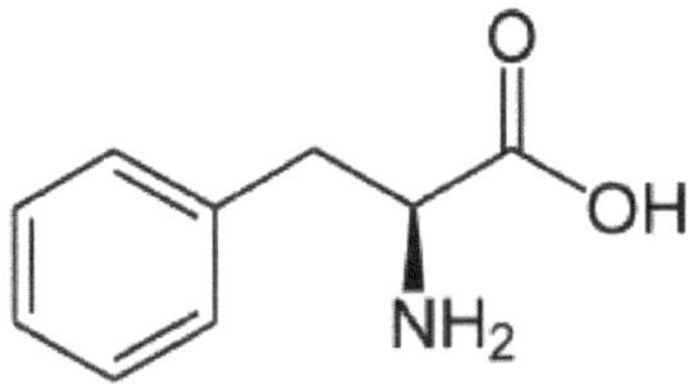

FIGURE 3.8 Structure of the amino acid phenylalanine.

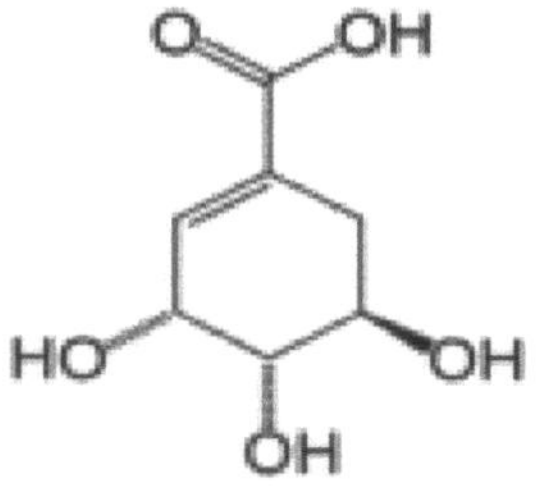

FIGURE 3.9   Structure of shikimic acid.

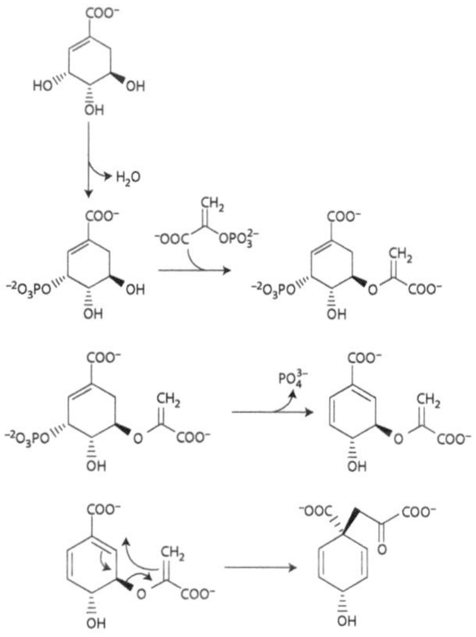

FIGURE 3.10   Illustration of the shikimate pathway to generate the amino acid, tyrosine.

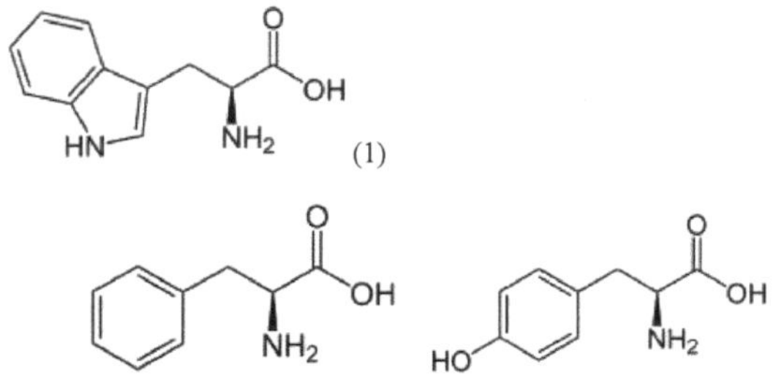

FIGURE 3.11   Aromatic amino acids: tryptophan (1), phenylalanine (2) and tyrosine (3).

SUSTAINABILITY AND SOCIETY'S CHALLENGES IN ENHANCING AGRICULTURAL PRODUCTION

Crop Yield and Green Issues

In crop production today, an urgent need exists to improve crop yield to feed a hungry world driven by an ever-expanding human population.

Over time in more developed parts of the world, nitrate and phosphate fertilizers have been applied liberally to the soil to stimulate quicker and thicker growth in cereal crops. To some extent this has been successful but at a cost to the environment. Over-reliance on chemical fertilizers produced in bulk by the inorganic chemical industry has led to diminished soil conditions as natural processes, which involve micro-organisms and improve fertility, are interrupted. The soil also becomes compacted under the weight of modern, heavy farm machinery. The run-off of rain water from fields pollutes water courses and lakes. Excess dissolved fertilizer promotes the growth of algal blooms, which reduce the light available to other green plants and drastically alters the balance of dissolved gases in the water such as oxygen and carbon dioxide, thereby choking fish and other invertebrates. These changes affect the food chain of other wild creatures (such as mammals, otters, and birds, herons and water rails), which depend on water life as a major food resource. Water quality for human use is also adversely affected.

Pesticides, insecticides and herbicides continue to be applied in large quantities to maintain a mono-culture in fields by protecting the tissues of crops and reducing or removing competition from other plants. These measures undoubtedly harm wildlife and may have an adverse impact on human health. Mammals, including human beings, are at the top of the food chain and consume cereals and other animals that have grown by eating grain. Some pesticides are made from chemicals that do not degrade naturally in the soil or general environment. There is a risk that minute quantities of these stubbornly resistant chemicals enter the food chain and can build up in human tissue over time with consequent and subtle effects on human health, which are not as yet fully understood.

Glyphosate is an herbicide that has been sold commercially worldwide—one well-known brand being Roundup®. Glyphosate is a non-selective herbicide that is unspecific in its action when applied as it has the ability to kill all plants. It is a systemic poison, which means that when it is absorbed through the leaves it is transferred to the growth points in the plant where it interferes with the shikimic pathway.

Glyphosate is an organophosphorus compound that is manufactured from an amino acid, glycine (NH_2CH_2COOH) and phosphorous acid.

Glyphosate has been used to suppress weeds in conjunction with crops that have been genetically modified to withstand it. This approach has economic value in that the practice of tilling the soil is very time consuming and uses up energy and also reduces soil compaction, rendering the soil more liable to erosion by wind and water. Since the shikimic pathway is not a part of mammalian metabolism it has been suggested that humans are immune from the effects of glyphosate. However, this may be an over simplification in that bacteria (which are single-celled plants) are present in vast numbers in the human intestinal tract. They are essential for good health in processing certain food items and in playing a part in the vital immune system. If glyphosate were to be admitted to the intestinal system, it could suppress the beneficial behavior of these valuable bacteria as suggested in a paper by Samsel and Seneff in 2013. As a result of concern, glyphosate is no longer sold in some developed countries.

Genetic Modification of Crops

Genetically modified crops have been created by man from natural strains of wheat. Genes have been artificially inserted into the genome of the cereal plant in order to selectively improve qualities valuable to man in the form of crop yield, good and consistent size in grain, flavor, resistance to drought and disease or in the length of the stalk, which if shorter is more harvestable, as it is less likely to blow over in windy, wet weather and involves less waste product in the form of straw. However, the long-term effect on the environment is as yet unknown. Despite great improvements in the efficiency of modern farm machinery, it is clearly impossible to collect every last scrap of grain from a field. Birds and the wind are bound to carry some of this grain away to the field boundary or beyond with unknown consequences FOR the environment. Unintended cross-fertilization of natural plants with genetically modified crops may well produce 'super' weeds, which are resistant to disease and the application of herbicides, thus creating the need for even more research and development. The energy and space requirements of genetically modified crops too are generally the same as those for naturally bred species of cereal.

ORGANIC COMPOUNDS CONTAINING PHOSPHORUS

Two of the fundamentally most important functions in plants (and animals) involve organic compounds containing phosphorus

- storage of genetic information in DNA and RNA and its retrieval for use
- metabolism, energy transfer within and between biological cells through molecules of ATP.

Deoxyribonucleic acid (DNA) is a huge polymer composed of two long chains made up from many monomers of nucleotide that coil around each other to form a double helix (figure). These polymers carry the genetic instructions for the growth, development, functioning and reproduction of all known organisms and many viruses.

Each monomer of nucleotide along a chain is composed of

- one of four cyclic amines with the properties of bases (cytosine [C], guanine [G], adenine [A] or thymine [T])

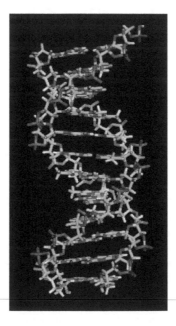

FIGURE 3.12   A model of the structure of a section of DNA showing the covalent links between the amine bases lying horizontally between the two spiraling strands.

Source: Zephyris. Creative Commons Attribution-Share Alike 3.0 Unported license

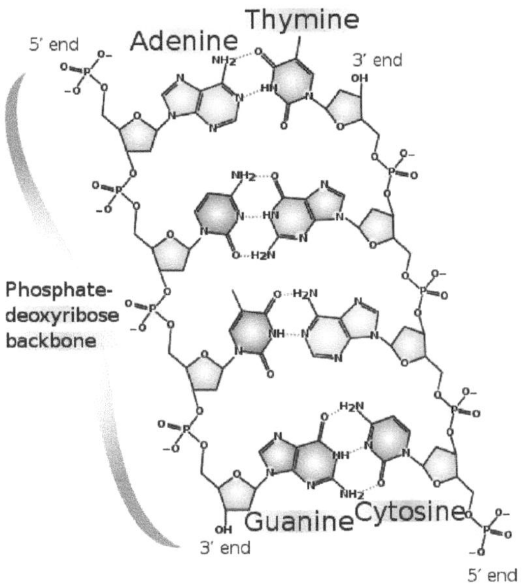

FIGURE 3.13 Chemical structure of a DNA helix showing the four amine bases and the phosphate and deoxyribose components of the 'backbone'.

Source: Madeleine Price Ball. Creative Commons CC0 1.0 Universal Public Domain Dedication

- a sugar, deoxyribose
- a phosphate group.

The amine bases provide cross-linking through hydrogen bonding while organophosphates along the polynucleotide provide a stiffening 'backbone' since the nucleotides are joined to one another along the chains by strong covalent bonds between the sugar of one nucleotide and the phosphate of the next, resulting in an alternating sugar-phosphate backbone.

The 5′ hydroxyl of each deoxyribose unit of sugar is replaced by a phosphate group forming a nucleotide monomer.

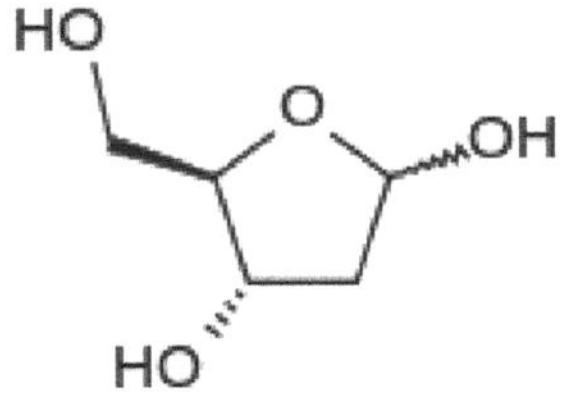

The spiral structure of the double helix of DNA is mainly due to the strength of the covalent links along each of the two chains of polynucleotide, which are formed by this phosphate ester providing a stiffening molecular 'backbone'.

DNA and ribonucleic acid (RNA), are nucleic acids, which are one the four major types of macromolecules that are essential to all known forms of life—the other three being proteins, lipids and polysaccharides.

A genetically modified crop is one in which the DNA located the cells has been altered or changed by removal or additions of sections. The biochemical engineering involved is achieved by physical methods or by the controlled use of certain bacteria so as to introduce traits into the plant crop that have economic value.

Food from genetically modified crops is widely consumed in the USA and Canada either directly or indirectly in processed material as a factor in soy protein, oils and flavorings. Here, food labeling does not need to draw attention to the use of genetically modified crops. In other parts of the world, notably in European countries, Australia and Japan, a degree of circumspection exists, where labeling is more explicit and in some instances food from genetically modified crops is prohibited from sale because it is not as yet considered to be proven to be safe.

Unless more land is brought into cultivation, which in itself has an impact on the habitat of wildlife and biodiversity, it is difficult to envisage how crop yields can continue to rise inexorably. Careful management and application of all of these techniques, coupled with research into more effective practices and materials, which can be demonstrated to be entirely harmless to man, continue to be required.

Questions

1. In modern farming, benefits should outweigh risks in order to justify the mass production of food items in a particular way. Give a balanced account of the pros and cons involved in the use of genetically modified crops.
2. Describe the chemical properties you would expect of phenylalanine given the functional groups present in the molecule.
3. Phenylalanine is a significant building block in the shikimic acid pathway through which secondary metabolites are formed in plants. Comment upon the differences between primary and secondary metabolites. Explain the particular interests that humankind has in both.
4. Figure 3.5 shows lysine, which is an example of an essential amino acid. Look up and draw the structures of the other essential amino acids: histidine, isoleucine, leucine, phenylalanine, threonine, tryptophan and valine.
5. A non-essential amino acid, serine (Figure 3.6), is shown in the chapter. Look up and draw structures for other non-essential amino acids: alanine, arginine, asparagine, aspartate, glutamic acid, glutamine, glycine and proline.

SUGGESTED FURTHER READING

J. Diamond. 1997. *Guns, Germs and Steel, A Short History of Everybody for the Last 13,000 Years*. Random House.

S. T. Grundas. 2003. *Wheat: The Crop. Encyclopedia of Food Sciences and Nutrition*, p. 6130. Elsevier Science Ltd.

J. F. Hancock. 2004. *Plant Evolution and the Origin of Crop Species*. CABI Publishing.

M. Hopf and D. Zohary. 2000. *Domestication of Plants in the Old World: The Origin and Spread of Cultivated Plants in West Asia, Europe, and the Nile Valley*, 3rd Edition, p. 38. Oxford University Press.

A. Kessel and N. Ben-Tal. 2012. *Introduction to Proteins: Structure, Function and Motion* Taylor and Francis.

A. L. Kolata. 2009. *Quinoa: Production, Consumption and Social Value in Historical Context*. Department of Anthropology, University of Chicago.

V. Kubyshkin, et al. 2019. *Anticipating Alien Cells with Alternative Genetic Codes: Away from the Alanine World!*. Curr Opin Biotechnol 60: 242–249. DOI: 10.1016/j.copbio.2019.05.006. PMID 31279217.

REFERENCES

DeepMind Technologies. 2020. *The AlphaFold Computer Program*. Nature 588: 203–204 and DOI: 10.1038/d41586-020-03348-4

S. Lev-Yadun, A. Gopher, and S. Abbo. 2000. *The Cradle of Agriculture*. Science 288: 1602–1603.

A. Samsel and S. Seneff. 2013. *Glyphosate Suppression of Enzymes and Amino Acid Biosynthesis*. Entropy 15(4): 1416–1463.

ASIAN STAPLE—RICE

Abstract: Rice feeds a greater proportion of the world's population than any other cereal crop. Cereal grain is rich in carbohydrates, which are plentiful in nature. Carbohydrates occur widely in both the plant and animal kingdoms, where they are present in supportive tissues and are a source of energy.

Organic Chemistry

- *aldehydes (alkanals)*
- *ketones (alkanones)*
- *carbohydrates*
- *cyclic compounds*
- *optical isomerism*
- *polymers.*

Context

- *monosaccharides (sugars, glucose and fructose) as natural building blocks*
- *disaccharide sugars, sucrose and lactose*
- *oligosaccharides.*

CARBOHYDRATES AND SACCHARIDES; OXYGEN IN THE ORGANIC RING

In organic chemistry, this very large class of substances is usually referred to in general terms as the carbohydrates, whereas in biochemistry they are typically known as the saccharides. The origin of the term saccharide is the Greek name for sugar, 'sacchar'. Incidentally, the name saccharin has been given to a man-made, artificial sweetener, which is neither a sugar nor a carbohydrate but has the functional group of an amine within a five-member carbon ring, which is integral to a benzene ring.

Carbohydrates are the most abundant class of organic compounds found in living organisms. The generic name, carbohydrate (*carbon hydrates*), arises from the observation that the molecular formula of this class of compounds is $C_n(H_2O)_n$, where *n* is typically a large number. Carbohydrates are produced by photosynthesis: the condensation of carbon dioxide requiring energy in the form of light and the green pigment, chlorophyll (see the chapter on 'Our World of Green Plants—Human Survival' in Part VII for more details).

$$nCO_2 + nH_2O + energy \rightarrow C_nH_{2n}O_n + nO_2$$

Monosaccharides, whether linear or cyclic, can be linked together in many different ways. Thus, a huge variety of individual carbohydrates and their isomers exist as disaccharides, oligosaccharides and polysaccharides.

Carbohydrates may be sub-divided broadly into two groups: sugars and non-sugars. Sugars, given the suffix, *ose*, are monosaccharides or disaccharides. Sugars are colorless, crystalline solids that are

TABLE 3.1

General Classification of Carbohydrates

Carbohydrates		
Sugars	*Non Sugars*	
Monosaccharides	Oligosaccharides	Polysaccharides
Pentoses Hexoses	Disaccharides (Tri and Tetra)	Homo Hetero
(Glucose, Fructose)	(Maltose, Sucrose)	(Cellulose, Starch)

soluble in water and usually have a sweet taste. In contrast, non-sugars are white, amorphous solids insoluble in water. Non-sugars can be broken down by chemical hydrolysis into their constituent sugars, which is why non-sugars are known as polysaccharides. If only one type of sugar is produced by hydrolysis, the non-sugar is known as a homo-polysaccharide, whereas if more than one different sugar is released it is described as a hetero-polysaccharide.

MONOSACCHARIDES

The simplest carbohydrates are called monosaccharides. They may be linear or may have either a six-membered or a five-membered ring structure. Representative examples are glucose and furanose, respectively, in Figure 3.14.

Even glucose, a monosaccharide, can exist in several isomeric forms (Figure 3.15), owing to the fact that the hydroxyl group can be in different configurations on each carbon atom in the ring.

DISACCHARIDES

When two monosaccharide molecules are joined together through a glycoside bond, they can form a disaccharide molecule. When more monosaccharide molecules link together, then a polymer called a polysaccharide is formed. Two disaccharides occurring widely in nature with the molecular formula $C_{12}H_{22}O_{11}$ are sucrose (Figure 3.16), the sugar of plants and lactose, the sugar of animals. The molecular structure of sucrose indicates how glucose is linked to fructose through a glycoside bond. Given that the systematic names of saccharides can be very long, arbitrary names remain in use for well-known substances.

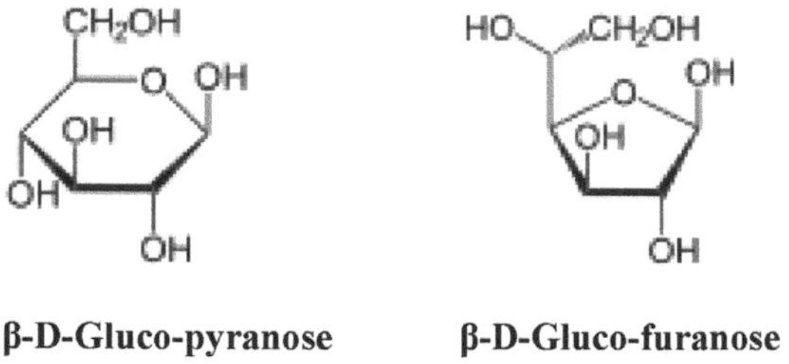

β-D-Gluco-pyranose   **β-D-Gluco-furanose**

FIGURE 3.14   Glucose and furanose.

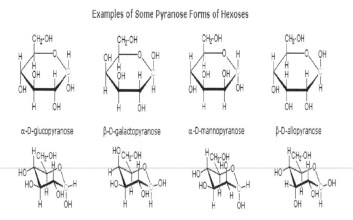

Examples of Some Pyranose Forms of Hexoses

α-D-glucopyranose   β-D-galactopyranose   α-D-mannopyranose   β-D-allopyranose

FIGURE 3.15   Examples of isomers of glucose.

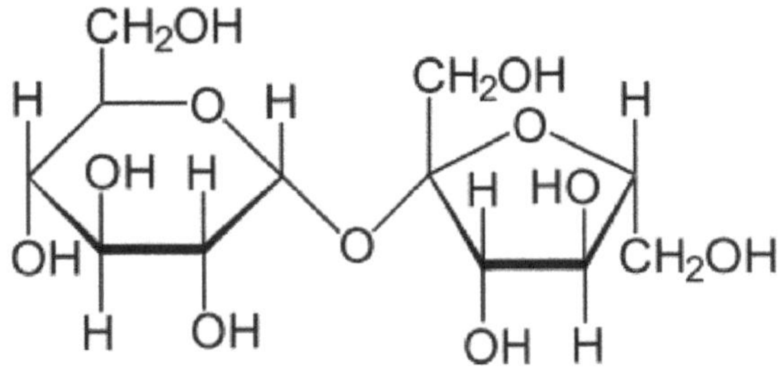

FIGURE 3.16 Structural formula of sucrose: glucose linked to fructose through a glycoside bond.

THE GLYCOSIDE LINK

The glycoside link is the name given to the ether-like bond involving the alpha carbon atom 1 in the six-member ring and the beta carbon atom 2 in the five-member ring.

Sucrose, commonly recognized as table sugar, is abundant as it is found universally in plants and in honey, which is derived by bees from plant material. Sucrose crystallizes readily from aqueous solution. Commercially, sucrose is obtained from the stem of the sugar cane grown in tropical lands and from the root of the sugar beet, cultivated in temperate parts of the world, notably northern Europe.

Lactose is the sugar found in the animal kingdom and is present in small concentrations in blood plasma and in the milk of mammals. It may be obtained commercially by evaporating whey, the aqueous liquor remaining as a by-product in the making of cheese.

OLIGOSACCHARIDES

Oligosaccharides are intermediate in size. Oligosaccharides are composed of several monosaccharides joined through glycoside links. As with other polysaccharides, the glycoside links can be hydrolyzed in the presence of enzymes or acid to yield the constituent monosaccharide units. A carbohydrate consisting of two to ten monosaccharide units with a defined structure is regarded as an oligosaccharide—anything larger would be referred to as a polysaccharide.

POLYSACCHARIDES

Very large polymeric molecules, made up from a high number and/or a great variety of monosaccharides, are known as polysaccharides (see the chapter on the 'Wonderful World of Wood'). It is easy to appreciate that polysaccharides can attain very high molecular weights of over 100,000 Da (see Glossary). Polysaccharides are a major source of metabolic energy, both for plants and for those animals that depend on plants for food. Polysaccharides are a component of the energy transport compound, ATP (see the Glossary and the chapter on the 'Chinese Cordyceps—Winter Worm, Summer Grass') and are found on the recognition sites of cell surfaces (see proteins and enzymes in the chapter on 'Wheat—Ancient and Modern').

CHEMICAL PROPERTIES OF CARBOHYDRATES—MONOSACCHARIDES

This book illustrates the chemical properties of carbohydrates through the relatively simple and common monosaccharides, glucose and fructose. The molecular formula of both is $C_6H_{12}O_6$. Although both sugars are good sources of energy, excess of glucose can be fatal to diabetic patients and excess of fructose in the human body can lead to health problems related to liver disease and to insulin resistance.

Whilst the name carbohydrate might suggest that these compounds are 'hydrates of carbon' and might behave as such, in reality they are more accurately described as poly-hydroxy aldehydes and poly-hydroxy ketones.

A linear molecule of glucose has the following form, although the aldehyde group could, of course, be present on any one of the carbon atoms along the chain.

$$CH_2 (OH).CH (OH).CH (OH).CH (OH).CH (OH).CHO$$

It is not surprising that this form of glucose displays the organic chemistry of the alkanals (or aldehydes) and alcohols. Given the presence of the aldehyde functional group, glucose is a reducing agent and is known as an aldohexose. It is a most important aldose.

In linear form, fructose has the structure

$$CH_2 (OH).CH (OH).CH (OH).CH (OH).CO.CH_2OH$$

While the ketone group could, of course, be present on any one of the carbon atoms along the chain, it is interesting to note that in naturally occurring fructose the carbonyl functional group is located on the second carbon atom. It is not surprising, therefore, that this form of fructose displays the organic chemistry of the alkones (or ketones) and the alcohols. Given the presence of the ketone functional group, fructose can react with hydrogen cyanide and amines. Fructose, therefore, is described as a ketohexose and is quite abundant, occurring in free form in many fruit juices and in honey.

It must be emphasized that the ring molecules of both glucose and fructose do not have either of these functional groups since the ring is closed by an ether linkage of carbon to oxygen to carbon. Also, it is worth pointing out that a number of isomers of each substance exist in the cyclic form of the molecule.

Under appropriate conditions, the hydroxyl in glucose or fructose can undergo reactions typical of a hydroxyl functional group;

- elimination of water to form the disaccharide, sucrose
- esterification with for example acetic anhydride to form an acetate and ultimately a penta-acetate
- reaction with either an aldehyde or a ketone involving neighboring hydroxyl groups to form a five-member cyclic acetal or a five-member cyclic ketal.

THE VALUE OF MONOSACCHARIDES, DISACCHARIDES AND POLYSACCHARIDES TO LIVING ORGANISMS

Monosaccharides are the major source of fuel for metabolism, being used both as an energy source (glucose being the most important in nature) and in biosynthesis. When monosaccharides are not immediately needed by many cells they are often converted into polysaccharides. In many animals, including humans, the polysaccharide in question is glycogen, which is retained in liver and muscle cells and can be readily converted back to glucose when energy is needed. In plants, starch is used for storage of surplus monosaccharides and for conversion back into energy. The most abundant polysaccharide is cellulose, which is the structural component of the cell wall of plants and many forms of algae. Fructose—or fruit sugar—is found in many plants. The sugar, ribose, is a component of RNA (see Glossary). Deoxyribose is a component of DNA (see Glossary).

HUMAN NUTRITION

Grain is a rich source of carbohydrates. Carbohydrates are a plentiful source of energy in living organisms although carbohydrates are not an essential component of human diet as human beings are able to obtain most of their energy requirement from protein and fats.

Organisms cannot metabolize every type of carbohydrate to yield energy, but glucose is an accessible source of energy. While many organisms can easily break down starch into glucose, most organisms cannot metabolize cellulose. However, cellulose can be metabolized by some bacteria.

Ruminants exploit microorganisms in their gut and termites farm microorganisms to process cellulose. Even though these complex polysaccharides are not very digestible, they do represent an important dietary element for human beings, known simply as dietary fiber. The 'rough' nature of dietary fiber aids digestion.

In food science, the term carbohydrate relates to food that is particularly rich in the complex carbohydrate, starch, such as cereals, bread and pasta or to simple carbohydrates, such as the sugars present in candies, jam and desserts. High levels of carbohydrates are often associated with highly processed foods or refined foods made from plants: sweets, cookies, candy, table sugar, honey, soft drinks, jam, bread, pasta and breakfast cereals. Lower amounts of carbohydrate are usually associated with unrefined foods, including brown rice and unrefined fruit.

Optical Isomerism

One of the most important consequences of the tetrahedral arrangement of single bonds around a carbon atom is *stereoisomerism*, namely, isomerism that results from the different arrangement in space of the atoms comprising a molecule.

There are two types of stereoisomerism: geometrical isomerism and optical isomerism. Geometrical isomerism is considered in Part II in *'Saving the Pacific Yew Tree'*.

Optical isomerism occurs when a molecule is not identical with its image when seen in a plane mirror. The two spatial arrangements of the molecule (the object and its image) cannot be superimposed just like gloves for the right and left hand. Optical isomers are known as enantiomers.

Consider the tetrahedral molecule of an imaginary substance derived from methane, Cwxyz. Since the molecule is asymmetric, two spatial arrangements are possible—two mirror images—two enantiomers.

No difference in physical properties can be envisaged because the enantiomers contain exactly the same atoms although they are arranged in a different spatial order. Thus it is found that the two enantiomers have the same melting points, boiling points, densities, vapor pressures etc. In one physical property, however, they do behave quite differently, which gives rise to the term optical isomerism, for the two enantiomers rotate the plane of polarised light equally but in opposite directions. The reason why this rotation of the plane of polarisation occurs is not fully understood. The angle of rotation is measured by placing the substance, whether in solid, liquid or gas state, in a polarimeter. An exploded view of a polarimeter is shown in Figure 3.17. Sinusoidal rays of light from the lamp oscillate in each direction in 360 degrees perpendicular to the line of travel. A device at 3, known as a polarizer, will only let light oscillating in a single plane pass through. Molecules with a chiral center, present in the sample at 6, are optically active and rotate plane polarized light either to the right (clockwise) or to the left (counter-clockwise) when that light passes through. A second polarizer at 7 must be rotated to allow plane polarized light through to the observer providing a measure of the angle of rotation.

The direction (clockwise or counterclockwise) and magnitude of the rotation reveals information about the sample's chiral properties such as the relative concentration of enantiomers present in the sample.

The enantiomer which rotates he plane of light in a clockwise direction when the observer is looking into the polarimeter is said to be *dextro*-rotatory, represented by the prefix *d* or (+). The enantiomer which rotates the plane in an anti-clockwise direction is said to be *laevo*-rotatory denoted by *l* or (-).

An equal mixture of the molecules of two enantiomers will be optically inactive. This is known as a *racemic mixture* or sometimes as a *racemate* and is denoted by *dl* or (+-). In a racemic mixture, optical inactivity is due to the balancing of the equal and opposite rotations of the two enantiomers. The polarimeter remains an invaluable tool for elucidating the composition of a mixture of different optical isomers.

While the chemical properties of enantiomers are identical because the same atoms or functional groups are present in both molecules and are bonded in the same way, the physiological action of each enantiomer can often be quite different. For example, the dextro forms of most amino acids can taste sweet while the laevo forms are bitter or tasteless. This fact is highly significant in the evaluation of the effectiveness of drugs in pharmacology.

Optical Isomerism in Sugars

Optical isomerism is common in sugars. While D-glucose is found naturally in plants, L-glucose does not occur in nature at all although it can be synthesized in the laboratory. D-glucose is the most abundant, naturally occurring monosaccharide. It is a building block for the disaccharides sucrose and lactose and many polysaccharides. D-glucose, in fact, is the only sugar unit in the non-sugar polysaccharides, cellulose and starch.

Since D-glucose is a basic chemical building block in the metabolism of many plant and animal organisms, greatly enhanced understanding in organic chemistry arose from accurate knowledge of its molecular structure. As a consequence, in 1902, Emil Fischer was awarded the Nobel Prize in chemistry for pioneering work in elucidating the structures of D-glucose and other sugars.

The glucose molecule can exist with its carbon atoms bound in an open-chain (acyclic) or in ring (cyclic; Figure 3.18). The prefixes D and L denote a chiral carbon atom. The chiral carbon atom is numbered 5 in the flattened presentation of the linear molecule in the Fischer projection and there is also one present at the first carbon atom in the ring in the alpha position.

OPTICAL ISOMERISM AND THE PHARMACEUTICAL INDUSTRY

If the enzymes in a living cell are harnessed to produce a drug that has a chiral molecule, natural plant metabolism will produce only one enantiomer. However, when the same organic compound is synthesized in the lab, it will instead contain an equal split of enantiomers—a racemic mixture. This represents a problem in the production of pharmaceutical drugs.

Drugs work by binding to receptor molecules in the body and thereby changing the course of chemical reactions. A drug molecule must have exactly the right shape to fit the site on the receptor molecule and only one enantiomer will do. The other enantiomer *might* fit a different receptor where it may have no effect at all or *may* cause harmful side effects.

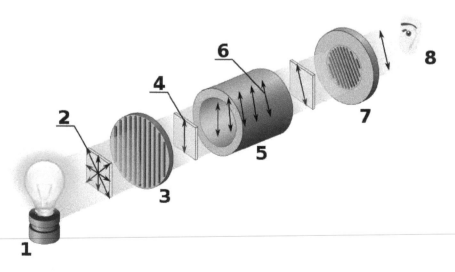

FIGURE 3.17   An exploded view of a polarimeter.

Source: Inkscape. Creative Commons Attribution-Sh are Alike 3.0 Unported license

FIGURE 3.18   (a) Fischer projection of D-glucose and (b) Alpha-D-Glucopyranose.

Synthetic chiral drugs usually have to be made so that they contain only one beneficially active enantiomer. The problem is that optical isomers, because they have very similar physical properties and chemical properties in general, are very difficult to separate so the synthetic production of single-enantiomer drugs is expensive. Exploiting organic chemistry in natural processes is clearly of commercial advantage provided that economy of scale can be achieved.

Questions

1. Explain what is meant by the term carbohydrate, and describe how this broad concept may be further classified into sub groups of compounds. Where do the following substances appear in this classification: glucose, sucrose, fructose, maltose, cellulose and starch? Show how oligosaccharides fit into the picture; give an example of one and explain its use or value.
2. Identify and systematically name the isomers of glucose.
3. Give the structure of each isomer of the linear molecules of glucose and fructose. Then, repeat the exercise for the cyclic forms of each compound.
4. Ethanol is being used increasingly as a fuel, especially in countries that have no or few reserves of petroleum. This has been true of Brazil in past years although off-shore reserves of petroleum have now been discovered and are being exploited. Sugar from sugar cane plantations is fermented to produce ethanol on an industrial scale and this is added to petrol or gas as bio-fuel. Explain why ethanol from this source is regarded as a carbon-neutral bio-fuel and give an account of any drawbacks.
5. Give an example of a chiral molecule and draw both enantiomers.
6. Describe and explain the effect of a racemic mixture on plane polarized light.
7. Ibuprofen and thalidomide are drugs that have optically active isomers. Explain why enantiomers arising in synthetically made drugs can be a serious problem for the pharmaceutical industry.

SUGGESTED FURTHER READING

N. G. Clark. 1964. *Modern Organic Chemistry*. Oxford University Press.
Innovia Films. 2010. *Performance Films for Beverages*. Innovia Limited.

CHINESE CORDYCEPS—WINTER WORM, SUMMER GRASS

Abstract: *'Winter Worm, Summer Grass'—the remarkable natural transformation of the mushroom, Cordyceps sinensis, which evolves from the caterpillar of a moth!*

Organic Chemistry

- *chemistry of fermentation*
- *redox reactions*
- *industrial production of bio-ethanol.*

Context

- *role of the fungus, Cordyceps sinensis*
- *adenosine triphosphate (ATP).*

THE LIFE CYCLE OF CORDYCEPS SINENSIS

The *Cordyceps* grows on the Tibetan Plateau. Through Chinese ingenuity it has become a health-giving, energy-boosting food. However, the mushroom has been collected almost to extinction.

The role of the fungus, *Cordyceps sinensis*, is explained. Curriculum content relates to the chemistry of fermentation and to redox reactions.

The fruiting body appears in summer as brownish-black 'blades' that are about 3–6cm long and are found amongst grass growing at an altitude of 3,000 meters on the Tibetan plateau (Figure 3.19). Apart from provinces of Sichuan and Yunnan in China, the mushroom is also found in Japan, Canada and Russia. The *Cordyceps* are fungi parasitic upon insect and arthropod larvae. Spores from the fungus present infect the larva. Then the spores develop into the thin, thread-like 'body' of the fungus, which is called the mycelium. The mycelium consumes the host larva, eventually killing and mummifying it. Fungi feed by absorption of the nutrients the filaments of mycelium find in their hosts. They are non-green plants without chlorophyll and cannot photosynthesize their own food.

One particular form of the mushroom, *Cordyceps sinensis*, is parasitic on the caterpillars of a moth that are colonized by the fungus underground. When host dies, the mycelium of the fungus produces a fruiting body above ground, which releases more spores to continue the life cycle (Figure 3.20).

THE PERCEIVED HEALTH BENEFITS OF CORDYCEPS SINENSIS

There are many tales in folklore about the fungus acting as an invigorating tonic. Reports suggest the *Cordyceps* was collected and made into medicinal teas by China's early rulers. In more recent times, its reputation gained even more prominence when the story broke that the national Chinese coach announced to the world the secret of the caterpillar fungus amid claims of a performance-enhancing nutrient taken by the Olympic Chinese athletes who broke world records at the 1993 China National Games in Beijing.

Modern technology is employed in order to examine whether ancient medicines possess any therapeutic effect. Since the *Cordyceps* is believed to enhance energy and improve performance, athletes can be tested using an ergometer (Figure 3.21). Using the 'ergometer, it is possible to measure increases in VO_{2max} (oxygen uptake). The idea is to measure heart rate, pulse, work output as measured by speed and distance on a stationary bicycle, and the athlete is connected to breathing apparatus that measures oxygen intake. Measurements indeed showed that the oxygen intake increased when taking the *Cordyceps*, and the athlete's capacity increased too.

The *Cordyceps* has enjoyed immense popularity in highly populated eastern China having being brought from areas of Tibet and western China to the international market. It is hailed as one of China's medical treasures.

In China, the wild fungus is sold as medicine or food. It can be found packaged in small bundles, tied with thread and often attached to the naturally myceliated larvae of the caterpillars. The fungus is eaten in soups or cooked with meats and is often administered to elderly patients recovering

FIGURE 3.19   Fruiting body of *Cordyceps sinensis.*

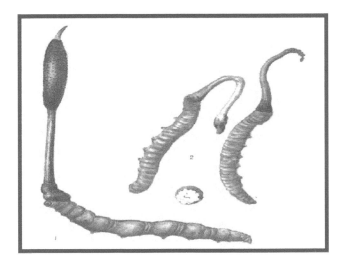

FIGURE 3.20   The fruiting body of *Cordyceps sinensis* (dark brown) protrudes from the earth and grows from the caterpillar of the moth, *Paecilomyces hepiali* chen (orange).

from illness. This seasonal, ancient Chinese extract has now become a popular fungal-based tonic reputed to address many illnesses and conditions. It is claimed that extracts of the fungus have been prescribed for illnesses ranging from headache to Asian flu and to cancer.

The fruiting body of the *Cordyceps sinensis* has become the source of one of the most sought after herbal extracts in the world and has been collected almost to the point of extinction on the Tibetan plateau.

Most of the world's supply of naturally produced fungus comes from China where an important industry has arisen providing income generation to relieve rural poverty. However, human consumption has been limited due to high price and short supply. Intensive research is being undertaken to generate sustainable supplies and meet spiraling demand. As an alternative, fungal strains from natural *Cordyceps sinensis* are isolated in an attempt to achieve a large-scale production by fermentation. The fungus is grown in fermentation cultures as pure mycelia in the liquid phase in China and in the solid state on grains in the western world.

FIGURE 3.21   An athlete using an ergometer.

WHAT IS FERMENTATION?

Fermentation is a process that can occur in nature or in the laboratory. It is a process that converts sugar to alcohol. Fermentation has been used by humans over millennia in the production of many foods and beverages.

Yeast is a single-celled fungus belonging to the phylum, *Ascomycota*, which reproduce by fission or budding. Yeast produces an enzyme that will convert an aqueous solution of sugar (glucose) into carbon dioxide and ethanol in a process known as fermentation.

Carbon dioxide gas bubbles out of the solution into the air leaving a mixture of ethanol and water. Ethanol can be separated from the mixture by fractional distillation.

Yeast is used to make beer and wine. An enzyme in yeast acts on the natural sugars in malt to make beer and grapes to make wine. When the concentration of alcohol reaches about 10–14% by volume the yeast dies and fermentation stops naturally, which is why wine is never any stronger unless it is fortified artificially by distillation.

Yeast of course is also used in the baking of bread. During baking, the carbon dioxide produced makes the bread rise by creating cavities and the alcohol evaporates.

Bacteria also promote fermentation. They are used to produce yoghurt and antibiotics such as penicillin. Fermentation also takes place in active, oxygen-starved muscle cells, leaving a residue of lactic acid which causes stiffness in muscle tissue if not flushed out in the bloodstream during a warm-down process after physical exercise.

The famous French microbiologist Louis Pasteur is often remembered for his insights into fermentation and its microbial causes. The science of fermentation is known as zymology.

ORGANIC CHEMISTRY OF FERMENTATION—THE ROLES OF ATP AND REDOX REACTIONS

Fermentation takes place in the absence of oxygen and is the biological cell's primary means of producing adenosine triphosphate (ATP), which is used to transfer energy.

ATP

Two of the fundamental and most important functions in plants (and animals) involve organic compounds containing phosphorus

- metabolism, energy transfer within and between biological cells through molecules of ATP
- storage of genetic information in DNA and RNA, which is examined in the chapter entitled 'Wheat—Ancient and Modern'.

ATP is often referred to as the 'molecular unit of intracellular energy transfer' because it is the organic compound through which energy is released to drive metabolism—the multiple and necessary processes of living cells. ATP consists of a base, a cyclic amine (adenine), attached to a sugar (ribose), which in turn is covalently bonded to a triphosphate group.

Incidentally, strong inorganic acids, such as phosphoric acid (H3PO4), will react readily with organic compounds such as amine bases to form amides and with the hydroxyl functional group of alcohols and sugars to form esters.

During metabolism, the adenine and sugar groups remain unchanged but the triphosphate group is converted to diphosphate and to monophosphate, the derivatives ADP and AMP respectively—and energy is released.

Other metabolic processes then regenerate ATP so the cycle can be repeated.

Redox reactions and ATP and ADP

Nicotinamide adenine dinucleotide (NAD) is an enzyme, a biochemical catalyst, which is found in all living cells. NAD exists in two forms, an oxidized state and a reduced state, abbreviated as NAD+ and NADH respectively.

The first step in fermentation is glycolysis, which involves breakdown of the energy store, glucose ($C_6H_{12}O_6$), into two ions of pyruvate (CH_3COCOO^-).

Two molecules of adenosine diphosphate (ADP) and two of phosphate (P_i) are then converted to two molecules of adenosine triphosphate (ATP) and two water molecules via a phosphorylation reaction (see Glossary). In the equation, ADP is adenosine diphosphate and P_i is phosphate.

$$C_6H_{12}O_6 + 2NAD^+ + 2ADP + 2P_i \rightarrow 2CH_3COCOO^- + 2NADH + 2ATP + 2H_2O + 2H^+$$

At the same time two molecules of the enzyme, nicotinamide adenine dinucleotide (NAD), are reduced to the enzyme, NADH. These reactions take place in the mitochondria of cells and they release energy. For more on redox reactions, see the Glossary and 'Tea: From Legend to Healthy Obsession' in Part IV.

INDUSTRIAL PRODUCTION OF BIO-ETHANOL

Fermentation by fungi is used on an industrial scale to produce a number of commercial products such as ethanol, citric acid, steroids and antibiotics. Fermentation must be carried out in the absence of air to make alcohol, otherwise ethanoic acid would be produced instead.

Ethanol can be made from algae by fermentation of the carbohydrates in the starch, which is stored by the plant as a food reserve. The ethanol produced can be used as a bio-fuel. Also, lipids in oil extracted on a large scale from algae can be made into bio-diesel fuel (see the chapter on 'Rediscovering the Traditional Grains of the Americas—Chia and Quinoa' in Part III).

Algae, unlike most green plants, have very thin cell walls with no structural tissue in the form of stems and leaves. Algae are therefore a good natural source of bio-ethanol because they are high in carbohydrates that the yeast can use but are low in cellulose.

The industrial process to produce ethanol includes the following stages:

- growing algae in an aqua culture
- harvesting the algae as a biomass
- introducing a yeast to the biomass to cause fermentation
- extraction of the resultant ethanol from the fermentation solution.

Algae are easy to cultivate in a photo-bioreactor either to fix CO_2 or to produce biomass. Photosynthesis is performed through chlorophyll in the green algae using sunlight as the energy source. Carbon dioxide is dispersed in aqueous solution in the reactor fluid to make it readily accessible to the algae. The equation describing photosynthesis is:

$$6CO_2 + 6H_2O = C_6H_{12}O_6 + 6O_2$$

Heat energy is released, which may be used productively for space heating or to power equipment. Fermentation by yeast of the carbohydrate extracted from the algae biomass releases CO_2, which can be stored and later recycled in the bioreactor to grow more algae thereby cutting costs.

Questions

1. Give three examples of redox reactions. Identify donors and acceptors and the changes in oxidation state for each one in your chosen reactions.
2. Ethanol can be made on a large scale for use as a fuel or a solvent from both renewable and non-renewable resources. Compare and contrast the benefits to society of the production of ethanol using fossil fuels with those arising from processes based on a renewable substrate.
3. Give examples of sources of industrially produced carbon dioxide and explain how this waste product may be usefully recycled.
4. Fermentation can take place under aerobic or anaerobic conditions. Explain the differences.
5. Why is the *Cordyceps sinensis* referred to sometimes as the *Winter Worm, Summer Grass?*

SUGGESTED FURTHER READING

C. Nappi. 2010. *Winter Worm Summer Grass, Cordyceps in Crossing Colonial Historiographies*, pp. 21–36. Edited by Digby et al. Cambridge Scholars Publishing.

Part IV

Beverages

INTRODUCTION

Well-known and popular, some non-alcoholic beverages are coffee, tea and cacao, each of which contains the stimulant caffeine. They have in common important roles in history and trade right up to the present day. Also included is a beverage derived from a less well-known plant called maca, which grows in Peru at elevations over 4,000 m. Maca is also used traditionally as a food.

A summary of the chemistry in Part IV follows.

Chapter	Organic Chemistry	Context
Tea	Phenolic ring	Catechins
	Electrophiles	Polyphenols (Flavonoids)
	Nucleophiles	
	Substitution reactions	
	Free radicals	
	Antioxidants	
Cocoa	Redox reactions	Flavonoids (polyphenols)
	Antioxidation	
	Free radicals	
	Human health	
Coffee	Isolation of caffeine	Caffeine
	Aromatic amines	
	Resonant structures	
	Zwitterions	
	Extraction and Purification techniques	
Maca	Secondary metabolites	Alkaloids
	Nitrogen in carbon rings	Indole
	Acid-base extraction	Indole alkaloids

DOI: 10.1201/9781032664927-4

TEA: FROM LEGEND TO HEALTHY OBSESSION!

Abstract: *From legend to healthy obsession; the popularity of drinking tea first emerged in Asia but now has conquered the rest of the world! Tea has a history that is inspiring and disheartening, uplifting and disturbing. Tea—and rituals and ceremonies associated with it—have been an inspiration for art and poetry for centuries but have also been the object of power and manipulation bringing both great pleasure and great pain to millions of people for many hundreds of years.*

Today, new scientific advances suggest important medicinal benefits and healing properties of green and black tea.

Organic Chemistry

- *phenol*
- *electrophilic substitution reactions and nucleophilic substitution reactions*
- *cis–trans isomerism*
- *chirality.*

Context

- *catechins.*

THE TEA PLANT

Tea (whether black, green or white) arises from a single species of plant, *Camellia sinensis*. Other beverages, popularly known as 'tea' prepared, for example from chamomile or mint, are more accurately described as tisanes.

Botanists have identified two distinct varieties. *C. sinensis var. sinensis* is indigenous to western China and has been cultivated there for nearly 2,000 years. *C. sinensis var. assam* is a relative newcomer to the world as it was only discovered in the Assam region of India in the 19th century. This variety is indigenous to a large geographic region: India, Myanmar, Thailand, Laos, Cambodia, Vietnam and southern China.

HISTORICAL NOTES ON TEA AS A BEVERAGE

No one really knows when the leaves of *Camellia sinensis* were first used to make tea. A legend dating from 3,000 BCE relates that a Chinese emperor, Shen Nung, was the first to taste tea. He was considered the father of Traditional Chinese Medicine in that he was said to have tasted and tested thousands of herbs to determine their possible usefulness. Archaeological evidence predating this legend suggests that tea was first consumed during the early Palaeolithic period (about 5,000 years ago).

Tea was originally used as a medicine and was considered so effective that, by the 4th century, it was an important part of Chinese life. It was used in attempts to cure a diversity of conditions including fatigue, rheumatic pain, kidney issues and breathing difficulties. Owing to the fact that successful processing methods had not yet been discovered, brewed tea originally had a bitter taste, which was masked by a variety of additives, including onion, ginger, salt, and oranges.

The art of processing tea evolved initially from direct use of the raw leaf to baking the leaves into a dried brick that could be carried great distances and used over a long time. The new processing method also resulted in a dramatically improved taste. Tea drinking then became very popular throughout China. During the T'ang Dynasty (618–907) appreciation of the visual arts, poetry, music and landscape gardening were all stimulated by trade in tea, both culturally and economically. Associated social rituals developed, resulting in teahouses and tea gardens springing up in cities and towns throughout the land.

By the time of the Song (Sung) dynasty, (~927 AD), tea had become one of China's most important items of trade. A new trade route promoted exchange between Tibet and the tea-growing regions of China. Records show that in one year alone some 20,000 Tibetan warhorses were traded for 34 million pounds of Chinese tea.

Although China was the only country at that time where tea was cultivated for export, appreciation of the beverage grew quickly abroad. By the 6th century, drinking tea was a part of daily life throughout Korea and Japan. In the 12th century, tea masters there had developed different ways of processing tea leaves. Instead of baking leaves into a brick, they were dried and powdered. Then boiling water was added and the brew was whipped with a bamboo whisk until foamy. Exposure of the brew to air in this way created a grassy, sweet taste, which is known today as Matcha tea.

Later in the 16th century in China, a new approach to infusing tea was tried, which involved drying the leaves carefully and then steeping them in hot water. A delicate, complex yet sweet flavor was released.

Eventually the British, amongst other nationalities, acquired a taste for this new flavor of tea. By 1675, tea could be found in food stores in England and, by the end of the 17th century, both black and green teas were being shipped to England from China in great quantities. In 1734, Thomas Twining opened the first teahouse in London, and others soon followed. By the end of that century, tea had become an essential part of British life. Drinking tea became a social occasion too. Late afternoon tea became an established custom throughout the country, first among the upper class and royalty but eventually spreading to lower social classes as well.

THE GROWING AND PROCESSING OF TEA

All true tea, green, black and white, is derived from the leaves of *Camellia sinensis*, an evergreen shrub of the *Theaceae* family. Unlike black tea, which has undergone some form of oxidation, green tea is harvested and carefully dried with little further process. Green tea is mainly consumed in the form of a brewed beverage. Successful tea cultivation requires moist, humid climates provided most ideally by the slopes of Northern India, Sri Lanka, Tibet and Southern China. Green tea is consumed predominantly in China, Japan, India and a number of countries in North Africa and the Middle East, whereas black tea is consumed predominantly in Western and some Asian countries and often taken with milk.

TEA AND POSSIBLE HEALTH BENEFITS

Historically, the medicinal use of green tea dates back to China 4,700 years ago. Drinking tea continues to be regarded in Asia as a generally healthy practice today. Scientific publications arising from clinical and epidemiologic studies tend to associate the health benefits of both black and green teas with the presence of organic compounds known as catechins, which are derivatives of phenol.

Research studies in both Asia and the West seem to indicate that drinking green tea contributes to the fight against many different kinds of cancer including those of the stomach, esophagus and the large intestine or colon. A correlation has been postulated between large consumption of green tea and low incidence of prostate and breast cancer in Asian countries, where green tea consumption is high. However, the influences of many other variables in lifestyle are a complication and mean that a definitive link between green tea consumption and beneficial effects upon cancer cannot as yet be established.

Indications are, however, more promising with regard to the prevention of skin tumors (melanomas) due to the effectiveness of catechins in absorbing harmful UVB radiation.

THE PROPERTIES OF PHENOL AND PHENOLS

Phenol, also known as carbolic acid, is an aromatic organic compound with the molecular formula C_6H_5OH having a phenyl group ($-C_6H_5$) bonded to a hydroxyl group ($-OH$; see also the chapter on 'Cocoa—Food of the Gods'). Phenol and its chemical derivatives (phenols) are key building blocks in the commercial production of polycarbonate materials, nylon, detergents, herbicides and numerous pharmaceutical drugs.

Phenol is also a building block of many natural products. The hydroxyl group in phenol may be replaced for example by a methyl or acetyl group or by a carboxyl group or by an ether linkage.

Phenols Compared to Alcohols

Although at first glance they may appear similar to alcohols, compounds based on phenol have unique distinguishing properties. Unlike in alcohols where the hydroxyl group is bound to a saturated carbon atom, it is significant that in phenol the hydroxyl group is attached to an unsaturated benzene ring. Phenol is quite soluble in water where partial dissociation takes place to give an acidic solution involving a negative phenoxide ion and a positive hydrogen ion. An explanation for the acidity of phenol is resonance stabilization by the aromatic ring of the phenoxide anion. Negative charge is delocalized over the benzene ring and the oxygen atom.

Electrophilic Substitution Reactions and Phenol

However, ease of electrophilic substitution (see Glossary) in the ortho and/or para positions (or the 2 and 4 carbon atoms) of the aromatic ring is a notable chemical property of phenol. Phenol will react with electrophilic substances such as nitric acid or concentrated sulfuric acid at room temperature to give in the latter case a mixture of ortho and para sulfonic acids. When, in turn, ortho and para sulfonic acids are fused with potassium hydroxide at 350 degrees Celsius, the corresponding dihydric phenols, catechol and quinol (hydroquinone), are produced. Their systematic formulae are $C_6H_4 (OH)_2$ (1, 2) and $C_6H_4 (OH)_2$ (1, 4) respectively. In this reaction, strongly nucleophilic hydroxyl anions substitute for the sulfonyl groups in the aromatic ring.

Interaction with Light in the Visible Part of the Spectrum of Electromagnetic Radiation

Light in the visible spectrum can interact with the de-localized electrons in the phenoxide ring, which gives rise to the variety of pigmentation in plants containing phenolic compounds. These aspects are explored more fully in the chapter entitled 'Colorful Chemistry; A Natural Palette of Plant Dyes and Pigments'.

CATECHINS—KEY PHENOLIC COMPOUNDS IN GREEN TEA

Catechin molecules present in tea contain four hydroxyl groups attached to two benzene rings which therefore have strong phenolic properties. Catechin is a flavanol (Figure 4.1), a natural phenol and an antioxidant, which belongs to the chemical family known as the flavonoids (see also the chapter on 'Cocoa—Food of the Gods' and 'Reversible Colors in Flowers, Berries and Fruit').

Catechin possesses two benzene rings (called the A- and B-rings) and a non-aromatic cycle (the C-ring) with a hydroxyl group on carbon 3. There are two chiral centers in the molecule on carbon

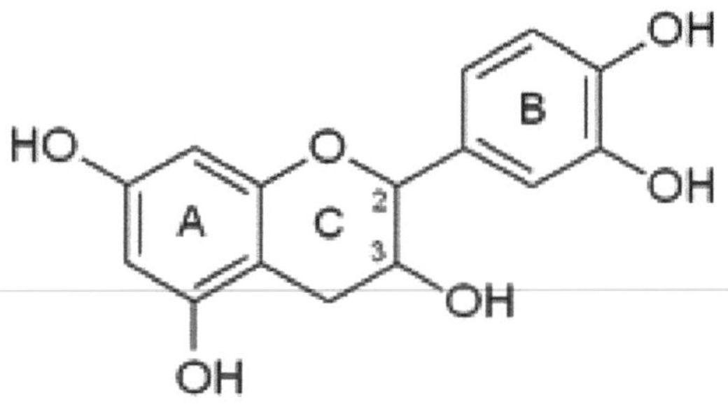

FIGURE 4.1   Catechin.

atoms 2 and 3 and so there are four diastereoisomers (see Glossary). The two isomers which are in the *trans* configuration and are called *catechin* while the other two isomers in the *cis* configuration are known as *epicatechin* (see also the chapter on the pacific yew tree for trans and cis isomerism).

Questions

1. Both phenol and ethanol contain the OH group. Describe reactions in which both substances behave similarly to one another and other reactions in which they behave quite differently.
2. Explain how water is important as a solvent in fostering partial dissociation of phenol.
3. Phenol is a building block of many natural products. The hydroxyl group in phenol may be replaced for example by a methyl or acetyl group or by a carboxyl group or by an ether linkage. However, ease of electrophilic substitution in the ortho and/or para positions (or the 2 and 4 carbon atoms) of the aromatic ring is a notable chemical property of phenol. Describe a theoretical model that adequately accounts for these properties of phenol.
4. Explain why concentrated rather than dilute sulfuric acid is necessary to allow electrophilic substitution in the benzene ring to take place.
5. Phenol is a tremendously important chemical used as a feedstock in the organic chemicals industry. Trichlorophenol, commonly known as TCP™, is a derivative of phenol that has medical applications due to its antiseptic properties. Describe three other diverse applications of products derived from phenol in the modern world.

SELECTED FURTHER READING

J. Blofeld. 1985. *The Chinese Art of Tea*. Shambhala.
A. MacFarlane and I. MacFarlane. 2003. *The Empire of Tea: The Remarkable History of the Plant That Took Over the World*. The Overlook Press.
L. C. Martin. 2007. *Tea. The Drink That Changed the World*. Tuttle Publishing.
R. Moxham. 2003. *Tea; Addiction, Exploitation, and Empire*. Carroll and Graf.

COCOA (CACAO)—FOOD OF THE GODS

Abstract: *Cocoa—food of the gods! Originating in South America and becoming the sacred beverage in Aztec and Mayan tradition, products of this modest tree of Amazon forests led to worldwide passion for chocolate and cocoa. Recent scientific data add medicinal importance to its value as a pleasant beverage for chocolate contains compounds known as flavonoids, which have been shown to have a wide range of pharmacological benefits and produce color in many plants.*

Organic Chemistry

- *properties of phenol*
- *polyphenolic compounds*
- *free radicals and antioxidants*
- *color arising from phenolic compounds.*

Context

- *flavonoids (polyphenols).*

ORIGINS

Everyone loves chocolate! But where does it come from? It was once prized by Aztec warriors and today by millions of people around the world. Furthermore, it may even be good for your health. Chocolate originates from cocoa. Today, cocoa consumption ranges from 0.1 kg/person/year in China to 11 kg/person/year in Ireland with the USA in the middle of this range at 5 kg/person/year.

The Mayans and Aztecs drank 'xocoatl' (cocoa) in an aqueous suspension—a bitter concoction made with chili peppers, corn mash, vanillin and other spices. Indeed, the term 'xocoatl' means 'bitter water'. Drinking a cocoa beverage before a long march or expedition was believed to increase energy and stamina.

European interest can be traced to the early 1500s, when Columbus engaged natives in the Gulf of Honduras, who gave him *xocoatl*. When sweetened with sugar, chocolate became popular throughout Europe. It was offered in fashionable drinking houses and valued for alleged aphrodisiac properties.

The chocolate or cocoa tree is commonly known as cacao to botanists. The scientific name of the chocolate tree (*Theobroma cacao L.*) literally means 'food of the gods'.

PROCESSING CACAO BEANS

Cacao beans (Figure 4.2) come from the pods found on cocoa trees growing in warm, humid places near the Equator.

FIGURE 4.2   Cocoa beans.

The main producing areas, Ghana, the Ivory Coast, Brazil and Nigeria are all perfect locations and production is increasing in Malaysia too.

Typically, cocoa seeds are fermented, dried and roasted during their processing into chocolate. For more on fermentation, see also the chapter on the 'Chinese Cordyceps—Winter Worm, Summer Grass' in Part III. Fermentation occurs in large piles of cocoa beans using the natural yeasts and bacteria that are present to produce some of the flavor precursors that we associate with chocolate. This processing creates a thick paste called cocoa liquor. Combination with cocoa butter (the fat component) and sugar creates dark chocolate and, when milk is added, milk chocolate. Chocolate contains fiber (most of which is lost with processing); minerals such as magnesium, copper and iron (providing a significant portion of the recommended daily allowance); the mono-saturated fatty acid, oleic acid and saturated fatty acids, mainly palmitic acid and stearic acid.

HISTORICAL NOTE—FIRST CHOCOLATE WAS SOLD AS A MEDICINE

In 1687 an English doctor and botanist, Sir Hans Sloane, was traveling in Jamaica where he tried their local chocolate drink, which was improved by adding milk, and this recipe was brought to England and first sold as a medicine. As more trade emerged, chocolate became increasingly popular, and to meet the new demand, cocoa plantations were developed in the West Indies, the Far East and Africa. However, the eating of chocolate was not in vogue until early Victorian times and then gradually spread across Europe as it became fashionable with the European royalty, the wealthy and the nobility. However, by the end of the 19th century, milk was added, leading to today's popular chocolate products.

COCOA AND CARDIAC HEALTH

In a very specific example of comparing yesteryear's and today's preparations we can observe two groups of Kuna Indians. A very low incidence of hypertension has been reported in Kuna Indian groups living on islands off the coast of Panama, who consume large amounts of unprocessed cocoa every day. The Kunas live an 'idyllic' island life and have much lower rates of cardiovascular diseases, hypertension, cancer and diabetes than populations of Kuna Indians who have migrated to the mainland Panama. Even though one may think that living on a tropical island would reduce blood pressure and stress levels, it appears that the Kunas on the mainland are as equally satisfied with their lives as the Kuna islanders. One explanation to account for differences in their health is that the Kuna islanders drink a minimally processed cocoa beverage throughout the day, whereas Kunas on the mainland drink a cocoa beverage that would resemble the more commonly and highly processed cocoa beverage available everywhere else.

COCOA AND DIABETES

The island population of Kuna Indians was also noted to have a lower incidence of type-2 diabetes. This disease results when the body no longer responds effectively to the insulin produced by the pancreas. The most important function of insulin is to direct the body to remove glucose from the blood and either store it in adipose (fat) tissue, store in the liver for future use or send it to muscle cells to provide energy for activity. Could there be a connection between cocoa consumption and the prevention of diabetes as well as cardiovascular diseases? There is now good evidence of a correlation between hypertension and insulin resistance. Insulin resistance refers to an inability to reduce blood glucose levels (hyperglycaemia) despite adequate production of insulin by the pancreas. Chronically high levels of glucose in the blood lead to oxidative stress, inflammation and eventually diabetes and cardiovascular disease. Insulin increases blood flow to micro capillaries in skeletal muscle, liver and adipose tissues and promotes cell surface receptors in these tissues to take up glucose.

Few would disagree that the seeds of this tropical tree, native to the upper Amazon rainforest, produce one of the most sought after food products, both today and in antiquity. Moreover, cocoa contains natural compounds that may reduce the incidence of diabetes and cardiovascular diseases. If, indeed, it can be shown that cacao helps prevent the insulin resistance and hypertension, which lead to cardiovascular disease and diabetes, then *Theobroma cacao* is truly both food and medicine of the gods!

CHEMICAL CONSTITUENTS OF COCOA

Cocoa seeds contain flavonoids, which are polyphenols. Polyphenols are powerful antioxidants that protect cells from damage by free-radicals and the resultant inflammation of tissues.

Free Radicals and Antioxidants

A free radical is an atom, molecule or ion that has unpaired valence electrons. This feature makes free radicals highly reactive toward other substances. Most free radicals are reasonably stable only at very low concentrations in inert media (such as a Noble gas) or in a vacuum.

Examples of free radicals are the hydroxyl radical (HO•)—a molecule that is one hydrogen atom short of a water molecule—and the oxygen atom, which is an intermediary in the formation of ozone in the upper atmosphere when ultraviolet light dissociates an oxygen molecule:

$$O_2 = 2O; O_2 + O = O_3$$

Free radicals may be created in a number of ways, including synthesis with very dilute or rarefied reagents, reactions at very low temperatures or the breakup of larger molecules. The latter can be brought about by ionizing radiation, heat, electrical discharges, electrolysis and chemical reactions.

Free radicals are intermediate stages in many chemical reactions. Free radicals and other oxygen-derived species are constantly generated in the body both by accident and during metabolic processes. The reactivity of different free radicals varies but some can cause severe damage to important biological molecules such as DNA (see Glossary) or those that form major organs such as the liver. Antioxidants offer some defense as they react with and minimize the formation of oxygen-derived free radicals. Consequently, antioxidants obtained through diet may be particularly important in helping us to stay healthier for longer.

Flavonoids

Flavonoids are polyphenolic compounds that also impart a purple color to the coating of the bean but are largely lost during fermentation, roasting and other processes leading eventually to the production of chocolate. Flavonoids are widely distributed in plants, fulfilling many functions. Flavonoids are the most important plant pigments for flower coloration, producing red, purple or blue pigmentation in petals to attract pollinators (see also the section on 'Colorful Chemistry; A Natural Palette of Plant Dyes and Pigments' in Part VII and, for more on flavonoids in particular, the chapter on 'Reversible Colors in Flowers, Berries and Fruit' also in Part VII). In higher plants, flavonoids are also involved in the symbiotic fixation of nitrogen from the air.

Most chocolate drinks contain about 25 mg or so of flavonoids. Chemical extracts reveal among them the presence of two common polyphenol compounds: theobromine and caffeine.

It is interesting to note that modern nutritional supplements, containing cocoa flavonoids or pure epicatechin, are often used by runners and body-builders who try to gain the same advantages that the Aztecs sought from the foamy cocoa drink. In the 18th century, concoctions were made by apothecaries and chemists who considered their cocoa blends as a kind of medicine. However, these forms of beverage were very different from today's chocolate. They contained several flavonoids, which are typically lost in today's processing and had no added sugars or milk. Food processing does affect the level of flavonoids remaining in the finished product and levels are widely variable between types of milk, dark and white chocolate and cocoa products.

Questions

1. The main producing areas of cocoa beans, Ghana, the Ivory Coast, Brazil and Nigeria, are all perfect locations and production is increasing in Malaysia too. What are the similarities and growing conditions needed in these different countries?
2. The process of fermentation has been introduced in the chapter on the 'Chinese Cordyceps—Winter Worm, Summer Grass' in Part III and reinforced this chapter on the processing of cocoa beans. Describe the process of fermentation in your own words.
3. Chemical extracts from cocoa reveal the presence of two alkaloids, theobromine and caffeine. Draw their chemical structures. Why are they considered as alkaloids?
4. What other well-known plant do you think contains theobromine and caffeine?
5. In a very specific example, a very low incidence of hypertension has been reported in Kuna Indian groups living on islands off the coast of Panama, who consume large amounts of unprocessed cocoa with high flavonol content every day. Why is monitoring hypertension important to our health?
6. Explain the difference between Type 1 and Type 2 diabetes.
7. What is insulin and what is its function in the human body?

SUGGESTED FURTHER READING

S. D. Coe and M. D. Coe. 1996. *The True History of Chocolate.* Thames and Hudson.
L. E. Grivetti and H.-Y. Shapiro, Eds. 2009. *Chocolate. History, Culture and Heritage.* John Wiley & Sons.

COFFEE—WAKE UP AND SMELL THE AROMA!

Abstract: Wake up and smell the aroma! With origins in East Africa, coffee has now become a global mega crop. This widely consumed beverage is recognized not only as a mild stimulant but also for health benefits.

Organic Chemistry

- *cyclic aromatic amines*
- *zwitterions*
- *the process of decaffeination of coffee.*

Context

- *caffeine*
- *nitrogen atoms in cyclic molecules.*

Early Use

The Kefficho people, living in the region known as Keffa in Ethiopia, are believed to be among the first to discover and recognize the energizing effect of an extract from coffee beans. The use of coffee is believed to have spread from Ethiopia to Egypt and Yemen. The earliest credible evidence of either coffee drinking or knowledge of the coffee bush appears in the middle of the 15th century in the Sufi monasteries of Yemen. By the 16th century, knowledge of coffee had reached the rest of the Middle East, Persia, Turkey and northern Africa.

Coffee was first imported to Europe through Venice. Trade wars were now on! The race among Europeans to secure coffee trees or fertile beans was eventually won by the Dutch. Largely through the efforts of the British East India Company, coffee became available in England no later than the 16th century. The first coffee house in England was opened in St. Michael's Alley in Cornhill, London.

The first coffee plantation in Brazil was established in 1727. Large tracts of rainforest were cleared for coffee plantations. However, by the 1800s, coffee went from being an elite indulgence to a drink for the masses. In the 19th and early 20th centuries, Brazil became the biggest producer of coffee.

Coffee and Caffeine

Coffee comes from the *Coffea* genus of flowering plants. Although caffeine is not responsible for the well-known aroma of coffee, caffeine is present in the seeds where it protects the seeds as a secondary metabolite (see Glossary) due to its toxicity to herbivores. One of the most popular of coffee-producing plants is *Coffea Arabica* (Figure 4.3).

The chemical formula of caffeine is $C_8H_{10}N_4O_2$. and its molecular structure is given in Figure 4.4.

Caffeine is weakly basic and is a white, colorless powder when anhydrous. The solubility of caffeine is 2g/100mL in water at room temperature, which increases significantly to 66g/100mL when it is added to boiling water.

Zwitterions

A zwitterion has both a positive and a negative charge at different positions within its structure. In other words, it is a dipolar ion. Amino acids (see the chapter on 'Foods of the Fertile Crescent—Ancient Wheat' in Part III) are well known for this property in that the molecules possess both basic and acidic functional groups in the form of an amine and a carboxylic acid. The six-member ring of caffeine contains two amide functional groups, which exist in resonance producing an

FIGURE 4.3   Coffea Arabica.

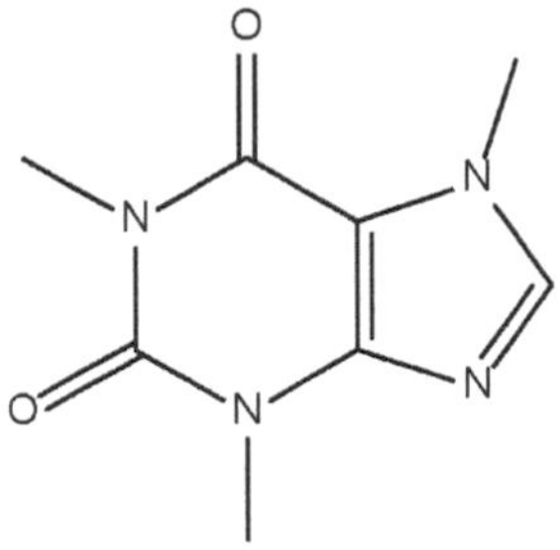

FIGURE 4.4   Structure of caffeine.

$$CH_3CH\ (NH_2)\ CO2H <=> CH_3CH\ (NH_3)^{(+)}\ CO2^{(-)}$$

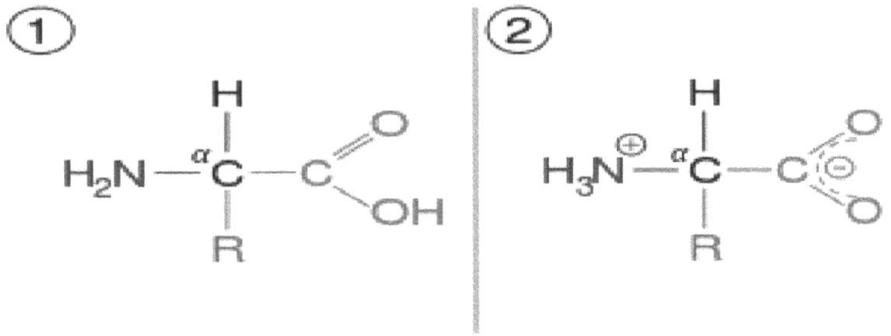

FIGURE 4.5   The zwitter ion forms of an amino acid.

intra-molecular separation of charge as in a zwitterion (see the diagrammatic representation of the structure in Figure 4.5.) The possibility of hydrogen bonding when caffeine is added to a polar solvent such as water helps to explain the moderate solubility of the compound.

Thus, in a similar manner, we can see the two resonant molecular structures for caffeine. The position of the equilibrium is related to the pH of the solution containing caffeine (Figure 4.6).

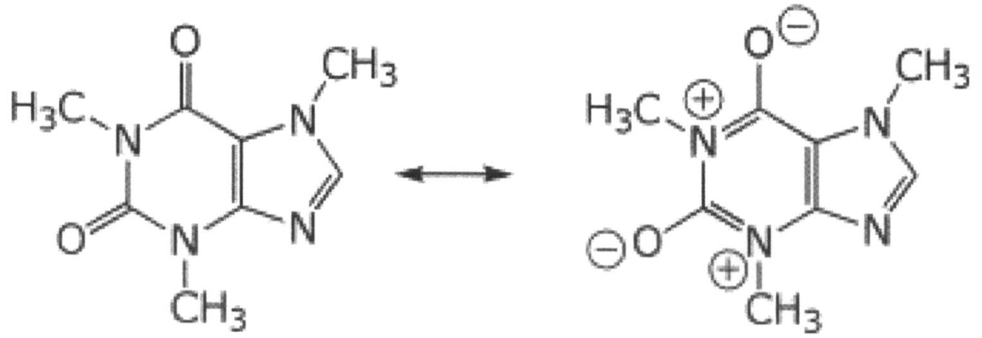

FIGURE 4.6   Resonant molecular structures of caffeine.

CYCLIC AROMATIC AMINES

A molecule of caffeine clearly possesses four nitrogen atoms in two joined, heterocyclic rings, which are aromatic in nature. Two of the nitrogen atoms are in the configuration of an amide, whilst the other two are formally in the relationship of a tertiary amine.

The chemistry of amines and amides is presented and reinforced in the chapter on 'Tobacco' in Part V. As amines are ultimately derivatives of ammonia, they contain a lone pair of electrons located on the nitrogen atom, which makes them basic compounds. However, the degree of basicity is markedly influenced by neighboring atoms and whether or not the nitrogen atom is incorporated into a heterocyclic, aromatic carbon ring as shown for caffeine. The aromatic structure decreases the basic properties of the compound due to the delocalization of electrons. This is an important point, which governs the fact that caffeine is only weakly basic. However, this property significantly affects the process of removal (or reduction in the level) of caffeine in beverages such as coffee and tea in the decaffeination process. As caffeine is weakly basic, it will dissolve in polar solvents such as water or ethanol.

HISTORY OF DECAFFEINATION

The first successful decaffeination was achieved in 1820 when the German chemist, Runge, analyzed the constituents of coffee to discover a possible link between drinking coffee and insomnia. There was a more significant breakthrough by Ludwig Roselius in 1903. He pretreated coffee beans with steam, which eventually became the basis for commercial production of decaffeinated coffee in the early 20th century.

THE PROCESS OF DECAFFEINATION OF COFFEE

There are several ways to remove caffeine from coffee and three methods are presented.

Extraction Procedure I: Solvent Extraction Using Water

Runge developed a commercial procedure that depends upon the solubility of caffeine in water that can be reproduced in the laboratory. Thus, clean coffee beans are first soaked in water. Caffeine dissolves along with some other compounds in low concentrations. The solution is then passed over charcoal. The charcoal absorbs and retains all but the molecules of caffeine, which can be recovered from the aqueous phase by standard laboratory techniques. The process is repeated with fresh charcoal. Eventually, the coffee beans are dried and are caffeine-free.

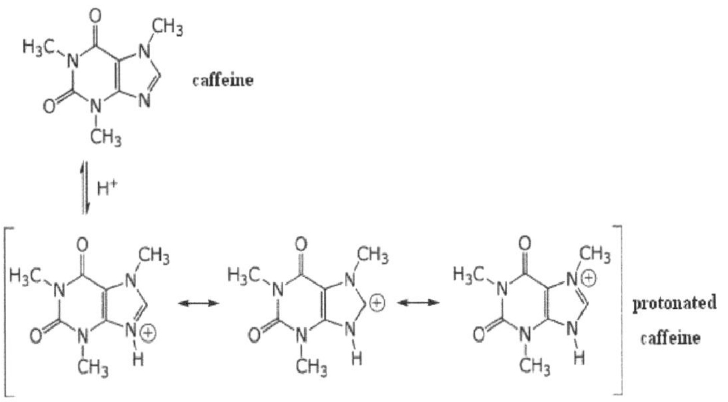

FIGURE 4.7 Addition of sodium carbonate converts the protonated form of caffeine to its free form.

Extraction Procedure II: Solvent Extraction Using Dichloromethane (DCM)

First, ground coffee in aqueous sodium carbonate is refluxed for 20 minutes; the mixture is filtered and cooled. The aqueous filtrate is partitioned into DCM and repeated several times to extract more caffeine. The addition of sodium carbonate, a weak base, converts the protonated form of caffeine, which is naturally present in coffee, to its normal, free caffeine form.

Extraction Procedure III: Supercritical Carbon Dioxide Extraction

There are two clear advantages to this method: elimination of the use of a flammable, toxic solvent and the caffeine being more easily removed from the final product. However, sophisticated equipment is required to sustain the high pressure and temperature required to maintain carbon dioxide in a supercritical fluid state (*SC-CO$_2$ for short, also see the Glossary for definition of the critical point*).

Supercritical carbon dioxide is an excellent non polar solvent for caffeine. In the extraction process, CO_2 is forced through the ground coffee beans at temperatures above 31.1°C and high pressures above 73 atmospheres. Under these conditions, CO_2 is in a supercritical fluid state when it has the properties of both a gas—allowing the solvent to penetrate deep into the beans—but also the properties of a liquid, whereby it can dissolve 97–99% of the caffeine. The solution of caffeine in supercritical SC-CO$_2$ is then sprayed with high-pressure water to remove the caffeine. The caffeine can be removed from aqueous solution by charcoal and refined by distillation or re-crystallization if required.

SC-CO$_2$, as a non-polar solvent, dissolves non-polar solutes from the mixture. The subsequent addition of a more polar solvent, water, dissolves the somewhat polar solute, caffeine. The careful selection and use of co-solvents in this way to partition a complex mixture enriches the extraction of a target solute, in this example caffeine, from a complex natural product without damaging the latter.

Multiple partitioning in a continuous industrial extraction process removes most of the caffeine. Coffee beans enter at the top of an extractor vessel with fresh CO_2 entering at the bottom. Recovery of caffeine is achieved with water in a separate absorption chamber. The process benefits from a pre-treatment step. The material first is soaked with ultra-pure water when hydrogen bonds linking caffeine to its natural matrix are ruptured. Swelling and bursting of the cell membrane also enhances diffusion of the solutes into the solvents.

The extract from the beans contains the compounds besides caffeine, which contribute to the flavor of coffee. Finally, the extract is dried, leaving decaffeinated coffee with its original flavor intact. The quality of recovered caffeine produced can reach a purity of greater than 94%, making it acceptable for use in the soft-drinks and pharmaceutical industries.

Questions

1. Which of the four nitrogen atoms in the caffeine molecule do you expect to be the most basic? Explain in terms of the delocalization of electrons over the two fused rings showing the possible different resonant structures of the molecule.
2. Explain fully in terms of molecular structure and chemical properties why caffeine, as an aromatic organic compound, is soluble in water at all.
3. Give a full account of how water is used in a supercritical fluid state in the industrial-scale generation of electrical power.
4. Previously, organic solvents such as hexane, benzene, chloroform and carbon tetrachloride have been typical choices for the extraction a solute by partitioning. Explain why the use of ethyl benzoate for this purpose is much more preferable in the laboratory.

SUGGESTED FURTHER READING

S. C. Chew. 1974. *The Crescent and the Rose*. Oxford University Press.
R. Cooper and G. Nicola. 2014. *Natural Products Chemistry; Sources, Separations and Structures*. CRC Press, Taylor and Francis.
Encyclopedia Britannica. 1954. *The Blessed Bean—History of Coffee*. Otis, McAllister & Co.
M. Pendergrast. 1999. *Uncommon Grounds: The History of Coffee and How It Transformed Our World*. Basic Books.

MACA FROM THE HIGH ANDES IN SOUTH AMERICA

Abstract: How did the Inca prosper at high altitude in the Andes? The answer may be the ancient staple food, maca. It looks like a little radish and is obtained from the plant, Lepidium meyenii.

Organic Chemistry

- *the properties of secondary amines located in a heterocyclic ring when compared with those of aromatic heterocyclic compounds containing nitrogen*
- *a simple practical technique—acid-base extraction*
- *steric effects in large molecules.*

Context

- *alkaloids*
- *indole as a building block in nature*
- *introducing indole alkaloids.*

THE MACA PLANT

Maca (*Lepidium meyenii*) grows in Peru at elevations over 4,000 m in an environment that would be exceptionally challenging for most plants: intense cold, intense sunlight and strong drying winds. It is a staple food and has been cultivated in the Andes for 1,500–2,000 years and may have contributed to the survival of a healthy indigenous human population in the High Andes.

MACA AS A BEVERAGE AND A FOOD

Once unknown in the Western world, maca is now being recognized well beyond the Andean region. The useful part of maca is found in the root. Maca can be consumed as a beverage. Juice is recovered from the root by boiling in water or by extraction with alcohol.

In foods, maca is eaten either baked or roasted and may also be prepared as a soup. Maca flour is being sold in health food stores as a more healthy option than conventional flour. In an increasingly inter-dependent world, cultural fusion in culinary practice has led to products such as spaghetti being made from maca flour.

FIGURE 4.8 The radish-like root of the maca.

The Medicinal Value of Maca

Traditionally, maca has been taken orally to relieve symptoms of anemia and chronic fatigue as it has the capability to enhance stamina, athletic performance and memory. Extracts of maca are also used to treat female hormone imbalance and menstrual irregularities and for enhancing fertility.

A study of the health status of adults 35–75 years old in the Peruvian central Andes involved comparison of those who used maca with those who did not. Maca was clearly associated with better health indicators: stronger bones (fewer fractures), reduced incidence of chronic mountain sickness, lower body mass index and lower blood pressure.

Chemical Composition of Maca

Dried maca root contains carbohydrates (60%); protein (10%); fiber (8.5%) and lipids (2%) and significant amounts of minerals including iron, calcium, copper, zinc and potassium.

Important alkaloids are present in trace quantities that are unique to maca and may be used to identify the plant from samples. Alkaloids are a large group of naturally occurring chemical compounds that are produced by a wide variety of organisms including bacteria, fungi, plants and animals. Alkaloids always contain at least one basic nitrogen atom. Generally, alkaloids (see *Glossary*) have biological effects on humans and animals.

Specifically, the tuber of the maca plant contains the alkaloid, (1R,3S)-1-methyltetrahydro-carboline-3-carboxylic acid, also known simply as a carboline. It is an indole alkaloid that is reported to exert influence on the human central nervous system. Its chemical structure is shown in Figure 4.9.

Indole

Indole is an important building block of many naturally occurring chemicals (see also the chapters on 'Africa's Gift to the World' and on 'Woad'). The indole building block is that group of atoms forming two fused rings on the left of the molecular diagram (see Figure 4.10). Indole is an aromatic,

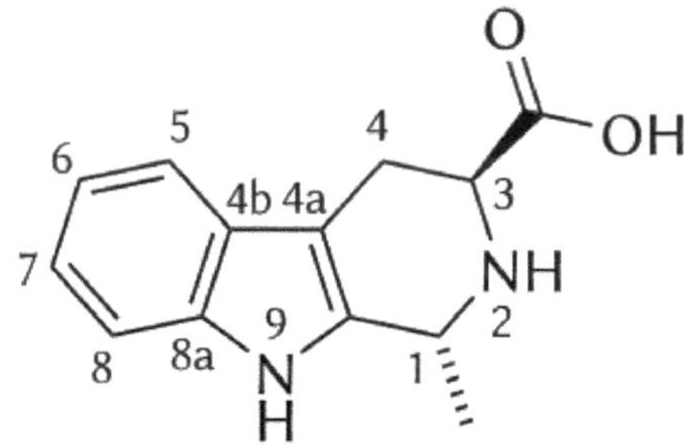

FIGURE 4.9   (1R, 3S)-1-methyltetrahydro-carboline-3-carboxylic acid.

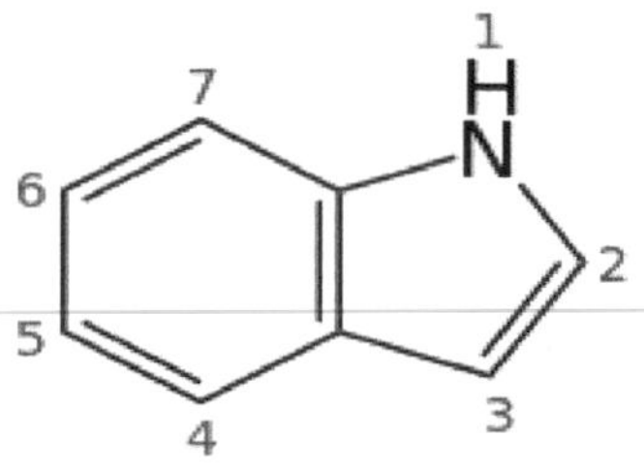

FIGURE 4.10   Indole structure showing the convention of numbering the carbon atoms.

heterocyclic organic compound. It has a bicyclic structure consisting of a six-membered benzene ring fused to a five-membered pyrrole ring. Pyrrole contains a nitrogen atom. For more on amines, amides and pyrrole refer to the chapter on 'Tobacco'.

Scientific interest in indole intensified when it is was realized that the indole building block is present in many important alkaloids.

Pure indole is a solid at room temperature. Indole is widely distributed in the natural environment and can be produced by a variety of bacteria, which are single-celled plants. Bacteria in the human gut produce indole as a degradation product of the amino acid, tryptophan, and so indole occurs naturally in human feces, producing that intense fecal odor.

The molecular structure of tryptophan contains the indole building block. The molecule is completed by an amino acid, derived from propionic acid, which is attached to the indole building block at the second carbon atom from the nitrogen atom (see Figure 3.11 for the molecular structure of tryptophan in the chapter 'Wheat—ancient and modern'). Though tryptophan is essential for human life, significantly, it cannot be synthesized in the body and therefore must be part of our diet. Essential amino acids, such as tryptophan, have been reviewed earlier in the chapter on 'Wheat—ancient and modern' to be found in Part III. Essential amino acids act as building blocks in the biosynthesis of proteins. Fortunately for the human race, tryptophan is a routine constituent in many foods, being plentiful in bananas, dates, milk products, chocolate, meat, fish, poultry, eggs, oats and peanuts.

KEY CHEMICAL PROPERTIES OF INDOLE

Electrophilic Substitution

By far the most reactive position on the aromatic indole molecule for electrophilic substitution is the C-3 carbon atom on the pyrrole ring, which is much more reactive than a carbon atom within a benzene ring.

It follows then that, since the pyrrole ring is by far the most reactive part of a molecule of indole, electrophilic substitution of the carbocyclic or benzene ring can only take place once the N-1, C-2, and C-3 positions have been substituted.

Much more on electrophilic substitution is presented in the chapter on 'Tea: From Legend to Healthy Obsession' in Part IV.

Indole and Bases

Initial impressions from molecular structures are deceptive as indole is quite unlike an amine. In fact, indole is hardly basic at all. This is entirely due to the fact that indole is an aromatic compound in which electron delocalization of the electron pair of the nitrogen atom over and below the plane of the two rings plays a large part in stabilizing the molecule. This situation is quite comparable to the properties of the similar heterocyclic aromatic compounds, pyrrole and pyridine. Consequently, only a very strong mineral acid, such as hydrochloric acid, is able to protonate the nitrogen atom of indole.

ACID-BASE EXTRACTION

Alkaloids are produced in nature by a large variety of organisms including bacteria, fungi, plants and animals. Alkaloids are a group of naturally occurring chemical compounds that almost always contain at least one basic nitrogen atom. They can be purified from crude extracts by acid-base extraction.

The theory behind acid-base extraction is that salts, which are ionic, tend to be water-soluble while neutral molecules tend not to be. The addition of a mineral acid to a mixture of an organic base and acid will result in the acidic functional group remaining un-charged, while the base will be protonated. If the organic acid, such as a carboxylic acid, is sufficiently strong, its self-ionization can be suppressed by the additional acid.

Conversely, the addition of a base to a mixture of an organic acid and base will result in the base remaining uncharged, while the acid loses hydrogen ions to give the corresponding salt. Once again, the self-ionization of a strong base is suppressed by the added base.

The acid-base extraction procedure can also be used to separate very weak acids from stronger acids and very weak bases from stronger bases.

Usually, the mixture is dissolved in a suitable solvent—such as dichloromethane or diethyl ether (ether)—and is poured into a separating funnel. An aqueous solution of the acid or base is added, and the pH of the aqueous phase is adjusted to bring the compound of interest into its required form. After shaking and allowing for phase separation, the phase containing the compound of interest is collected. The procedure is then repeated with this phase at the opposite pH range. The process can be repeated to increase the separation. It is often convenient to have the compound dissolved in the organic phase after the last step so that the evaporation of the solvent yields the product.

Acid-base extraction is also covered in the chapters on 'Africa's Gift to the World; the Madagascan Periwinkle' and 'Cocaine'.

PURIFICATION OF INDOLE ALKALOIDS BY ACID-BASE EXTRACTION

Since the molecules of indole alkaloids contain at least one, weakly basic nitrogen atom, the indole alkaloids can be purified from crude plant extracts by acid-base extraction.

The carbolines present in the maca plant are indole alkaloids that are formed by plants through a condensation reaction involving amino acids and reducing sugars (aldoses). Significantly for the purposes of acid-base extraction, carboline has the functional group of a secondary amine at the number 2 carbon atom.

INDOLE ALKALOIDS

The indole alkaloids form one of the largest classes of alkaloids with more than 4,100 different compounds known. There are two types of indole alkaloid—those that also include an isoprene building block and those that do not. For example, the indole alkaloid found in maca is an indole alkaloid without a terpenoid group.

The physiological action of certain indole alkaloids on human beings, animals and birds has been well known down the ages. The group of indole alkaloids containing terpenoids has some interesting members, described later!

Strychnine

The French chemist Pelletier is revered as the founding father of alkaloid chemistry. The indole terpenoid alkaloid, strychnine, was isolated as early as 1818 by Pelletier and Caventou.

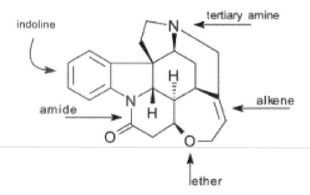

FIGURE 4.11 Molecular structure of strychnine showing the indole building block.

They used extracts from the plants of the *Strychnus* genus as source material. *Strychnus* is a flowering tropical plant belonging to the family, Loganiaceae or Strychnaceae, which also includes trees and lianas. The roots, stems and leaves of these plants are well known to indigenous peoples as sources of poisonous compounds, such as strychnine and curare, which they used efficiently in hunting (as discussed later in this chapter).

The strychnine molecule has a number of functional groups, one of which, the ether link, is not treated extensively in this book, primarily because it is relatively uncommon in naturally occurring compounds and it is not very reactive, given the stability of the carbon-oxygen single bond. This subject is also treated in the chapter entitled 'Attacking Malaria: a South American Treasure (but not gold) and a Chinese Miracle'. The relative stability of the carbon-oxygen single bond is in marked contrast to the reactivity of the carbon-oxygen double bond, which is somewhat strained. The latter functional group features in the extensive organic chemistry of the aldehydes, ketones, carboxylic acids and esters covered extensively in the chapters on 'Asian Staple—Rice', 'A Plant from the East Indies, Camphor', 'A Steroid in Your Garden', 'Morphine—A Double-Edged Sword' and 'Europe Solves a Headache'.

Ergotamine

It has been recognized for a long time that human consumption of contaminated grains of the cereal rye, affected by the ergot fungus, which is parasitic upon it, causes death by poisoning. The active chemical, an indole alkaloid called ergotamine, (Figure 4.12) was isolated from extracts in 1918.

LSD

Lysergic acid diethylamide, known in everyday language as LSD, (Figure 4.13) was first synthesized from naturally occurring ergotamine in 1938. LSD is a prohibited recreational substance in many countries of the world, as it has pronounced psychedelic properties. LSD is a large molecule made up from two of the most common building blocks found in nature—an indole and a terpene.

Curare

The name curare comes from the South American Indian word *ourare*, meaning arrow poison. Curare has been used with arrows or darts by South American natives to hunt animals (Figures 4.15 and 4.16). They are able to eat the animals subsequently without any untoward effects. The naturally occurring, active extract, tubocurarine, was first isolated in London by H. J. King in 1935 from a sample obtained from the large, climbing liana, *Chondodendum tomentosum*, native to the rainforest in South America. He also identified the structure of the molecule.

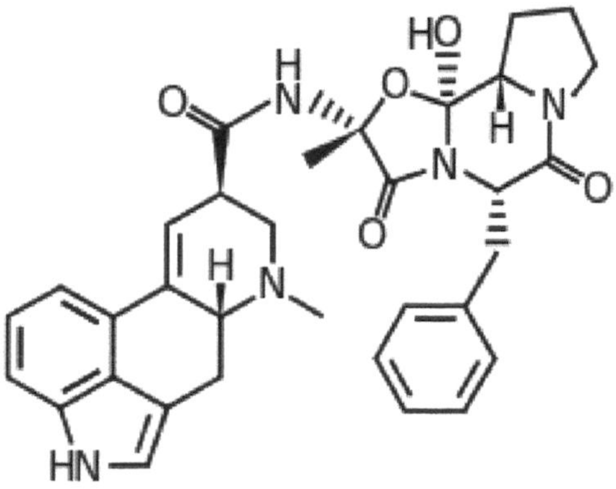

FIGURE 4.12   Structure of ergotamine.

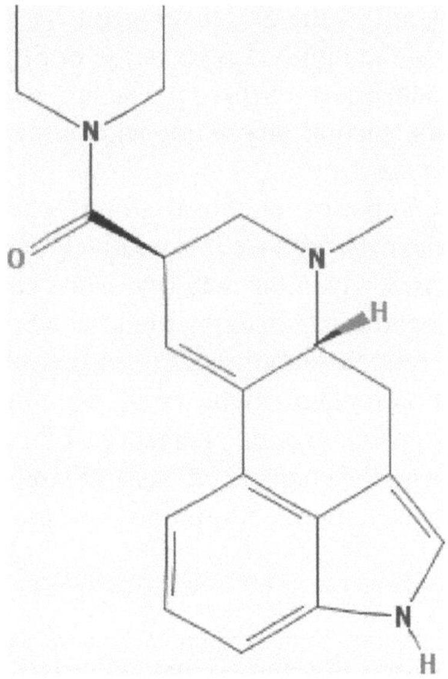

FIGURE 4.13 Structure of LSD.

FIGURE 4.14 Amazonian Yagua tribesmen in Iquitos, Peru can kill a monkey 30 m away with a blowpipe firing darts tipped with curare.

Source: with permission under terms of GNU Free Documentation License

Tubocurarine is a toxic alkaloid and skeletal muscle relaxant, known as a long-duration antagonist for the nicotinic acetylcholine receptor. In biology, a chemical compound, such as acetylcholine, which binds to a receptor and activates the receptor producing a biological response, is given a special name. The compound is known as an agonist. A compound such as tubocurarine, which blocks this process, is known as an antagonist.

FIGURE 4.15 Blowpipe darts are used in the Peruvian Amazon, tipped with a deadly poison (curare) from one of the Amazonian rainforest plants.

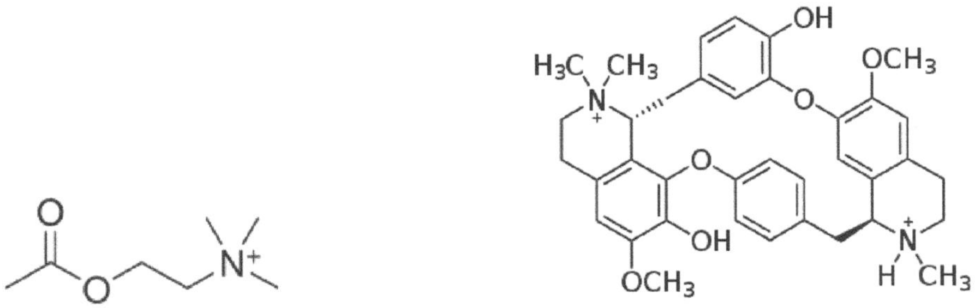

FIGURE 4.16 Acetylcholine, a neuro-transmitter and tubocurarine, a component of curare.

Acetylcholine is an organic substance that functions as the neuro-transmitter activating neuro-receptors at neuro-muscular junctions. You can see from Figure 4.16 that acetylcholine is an ester formed from acetic acid and an amino alcohol called choline.

Acetylcholine is present in the peripheral nervous systems of animals where it activates the neurons, which control the movement of muscles, notably those attached to the skeleton. Acetylcholine is also known to occur in the brain too where it is believed to influence learning, memory and mood and could be linked with the memory difficulties experienced by sufferers of Alzheimer's disease.

A neuro-receptor found in the muscle tissues of animals is nicotinic acetylcholine, which is activated by acetylcholine. The binding of the acetylcholine ion to that of the neuro-receptor is reversible. The 'firing' of the neurons attached to muscle tissues causes the muscles to contract followed by relaxation when the reaction is reversed.

However, the tubocurarine ion in curare can also bind to these neuro-receptors, but it does so irreversibly and therefore causes paralysis in muscle tissue. The diaphragm, which evacuates the lungs of air-breathing animals, is a large muscle that is also affected as curare from the tip of the dart is carried around the bloodstream. In effect, the animal dies from suffocation.

Even a cursory comparison of the molecular structures of acetylcholine and tubocuranine reveals why curare is so effective. Tubocurarine blocks the access of acetylcholine by binding to the neuro-receptor. Tubocurarine is an alkaloid with a molecule consisting of a cyclic system with a quaternary ammonium ion. Acetylcholine has a much simpler molecule as there is no cyclic system, but, significantly, it does also contain a quaternary ammonium ion. This shared feature means that curare alkaloids can bind readily to the active site of receptors for acetylcholine, which are located at the junctions between neurons and muscle tissue. Nerve impulses cannot be sent to the muscles that operate the skeleton, effectively paralyzing the animal.

The nicotinic neuro-receptors themselves, present at the junction of nerve tissue and muscle tissue, are very large molecules, measuring 290kDa. The molecules are tubular in shape with five openings at one end. Amine functional groups, which are electron rich, are exposed in this open structure. Ligands can be formed with this large molecule when specific molecules or ions that are electrophilic, such as acetylcholine or tubocurarine, bind with it. More on the chemical binding of ligands is to be found in the chapter entitled 'Our World of Green Plants—Human Survival' in Part VII.

Questions

1. What functional groups and common naturally occurring chemical building blocks are present in (1R, 3S)-1-methyltetrahydro-carboline-3-carboxylic acid? Describe the chemical properties you would expect the compound to have.
2. Explain why the substance (1R, 3S)-1-methyltetrahydro-carboline-3-carboxylic acid may be separated from crude chemical samples of the maca plant by acid-base extraction.
3. Explain fully the reasons why the nitrogen atom in the indole building block is not basic.
4. Identify all of the functional groups present in strychnine and describe the range of organic chemistry that you expect the compound might show.

SUGGESTED FURTHER READING

M. E. Moseley. 2001. *The Incas and Their Forebears: The Archaeology of Peru.* Thames and Hudson Publishers.
National Research Council. 1989. *Lost Crops of the Incas: Little-Known Plants of the Andes with Promise for Worldwide Cultivation*, p. 57. National Academy Press.
D. Seigler. 2001. *Plant Secondary Metabolism.* Springer Publishers.

REFERENCE

H. J. King. 1935. *Extraction of Tubocurarine from the Liana, Chondodendum Tomentosum, and Identification of the Structure of the Molecule.* Chem Soc 57: 1381 and Nature 135: 469.

Part V

Euphorics

INTRODUCTION

Certain drugs are known as euphorics because of their impact on the human brain. Plants containing substances that generate euphoria have played an important role in the culture of humankind. As source plants and their extracts were gradually 'discovered' by explorers, trade in 'new' drugs became embedded in the global economy and linked inextricably with colonial expansion.

Examples of controlled euphoric drugs are opium from the poppy plant, cocaine from the coca leaf and marijuana from cannabis. If taken properly under medical supervision, these drugs are of great benefit to humankind. However, we wish to make it clear that we do not condone their use for recreational purposes.

There are also 'soft' drugs, such as nicotine in tobacco and caffeine in tea and coffee, which have a degree of social acceptance.

SUGGESTED FURTHER READING

M. Jay. 2011. *Emperors of Dreams. Drugs in the Nineteenth Century*. Dedalus.
    A summary of the chemistry in Part V follows.

Chapter	Organic Chemistry	Context
Asian Poppy	Esters, Alcohols and Acids	Morphine, Codeine
	Acetylation	Heroin
Cannabis	Isomers	Cannabinoids
	Polymerization from	Terpene
	Elimination reactions	Isoprene rule
Coca	Cyclic amines	Cocaine
	Salt formation	Alkaloids
Tobacco	Amines, Amides	Nicotine—an alkaloid
	Heterocyclic aromatics	Pyridine ring
	Amines as building blocks	Pyrrole ring

DOI: 10.1201/9781032664927-5

MORPHINE: A DOUBLE-EDGED SWORD

Abstract: *The poppy has been traded through Asia and Arabia for centuries and continues to have an impact on underground commerce and societies. The poppy yields one of the most important and effective narcotics used in medicine today. Morphine is a legal drug used by medical doctors to relieve severe or agonizing pain and suffering. The drug acts directly on the central nervous system.*

Organic Chemistry

- *carboxylic acids*
- *alcohols*
- *esters*
- *acetylation.*

Context

- *the alkaloids: morphine, codeine and heroin.*

THE DOUBLE-EDGED SWORD

Every day the local hospital receives many patients. Patients requiring critical attention are always admitted first through the emergency ward. On one evening, two teenagers arrived separately. Both required immediate care but for different medical needs, although you will soon discover that their stories are very much linked. The first teenager, Jim, had had a serious bike accident and it appeared that his leg was fractured. He was feeling so much pain, the doctors needed to administer by injection several small doses of a pain killing drug called morphine. After the surgery to reset the fractured bone, he recovered well and the doctor carefully prescribed some more morphine to keep him comfortable and reduce the pain. Even though Jim felt better and was grateful for all the painkillers, he worried whether he could become addicted to the morphine that he had been receiving for the past couple of days. The doctors assured him that the likelihood of this was extremely low given his short exposure to the drug and that he would make a complete recovery. A week later, Jim left the hospital with a big plaster cast on his leg—ready for school friends to scribe get-well messages. The second teenager, Jack, entered the emergency room a few hours later. He had been found comatose. The doctor soon discovered he was suffering from an overdose of drugs—in this case heroin—but the doctors and nurses were slowly able to revive him. When he began to recover, Jack told the doctor how he had started on recreational drugs and began to smoke heroin. Then he learned how to inject himself with the heroin with much stronger effect on his body and mind. Jack explained to the doctor how he enjoyed it at first, but became addicted and continually craved more. Without the drug, he felt miserable. However, to continue buying the drugs, he needed to steal money and do other illegal things. The doctor told Jack that withdrawal from this powerful drug is a long and painful process, needing much medical patience, time and support, but there was good chance of recovery, if Jack was willing to work with the doctors and did not take any more heroin. Not all addicts are able to 'stay clean', but for Jack that night in the emergency ward became his first big step on the road to recovery.

Morphine, opium and heroin are all related chemicals and all have the high potential for addiction. Tolerance and psychological dependence develop rapidly, although physiological dependence may take several months to develop.

Morphine and opium are both classified as a narcotic—a drug to dull the senses. This means they have pain-killing properties and euphoric and hallucinatory effects. Morphine deadens pain, produces elation, induces sleep, and reduces stress. Opium induces gentle, subtle, dream-like hallucinations very different from the fierce and unpredictable weirdness of the psychedelic drug known as LSD.

A SHORT HISTORY OF OPIUM

Ancient peoples either ate parts of the poppy flower or converted them into liquids to drink. By the 7th century, the Turkish and Islamic cultures of western Asia had discovered that the most powerful medicinal effects could be obtained by igniting and smoking the poppy's congealed juices, and the habit spread. The widespread use of opium in China dates to tobacco-smoking in pipes introduced by the Dutch from Java in the 17th century. Whereas the Javanese ordinarily ate opium, the Chinese smoked it. The Chinese mixed opium with tobacco, two products traded by the Dutch. Pipe-smoking was adopted throughout the region, which resulted in increased opium-smoking, both with and without tobacco.

The great renaissance scientist, Paracelsus (1490–1541), concocted laudanum ('something to be praised') by extracting opium into brandy (ethanol), thus producing, in effect, a tincture of morphine. Laudanum was historically used to treat a variety of ailments but its principal use was as an analgesic and cough suppressant. Laudanum can be habit-forming. As their opioid tolerance increased, so did users' consumption of tinctures. By the 19th century, vials of laudanum and raw opium were freely available at any English pharmacy or grocery store. Until the early 20th century, laudanum was sold without a prescription and was a constituent of many patent medicines. During these times, opium was viewed as a medicine, not a drug of abuse. Also, the chemists and physicians most actively investigating the properties of opium were also its dedicated consumers, and this may have colored their judgment. Today, laudanum is recognized as addictive and is strictly regulated and controlled throughout most of the world.

In 1907, the British phased out India's opium export to China. At the time, they realized the social consequences of heroin, which is a derivative of opium, and they had lost out to Bayer in heroin production, and China began poppy field eradication. But even though treaties were enacted calling for countries to ban illegal trade of opium, criminal syndicates and illegal trafficking of opium became prevalent. But the use of opium for medicine was legal.

In North America, the initial history of poppy was somewhat more peaceful. During the first few centuries of European settlement, opium poppies were widely cultivated, and early settlers dissolved the resin in whisky to relieve coughs, aches and pains.

As well as being used for opium production, poppy seeds were used as food. The plant produces lots of small black seeds and they are used as a common garnish on rolls. Poppy seeds can also be ground into flour; used in salad-dressings; added to sauces as flavoring or thickening-agents and the oil can be expressed and used in cooking. Poppy heads are infused to make a traditional sedative drink.

Doctors had long hunted for effective ways to administer drugs without ingesting them. Taken orally, opium is liable to cause unpleasant gastric side effects. The development of the hypodermic syringe in the mid-19th century allowed the injection of pure morphine. Morphine use became rampant in the United States after its extensive use by injured soldiers on both sides of the Civil War. In late 19th-century America, opiates were cheap, legal and abundant.

Only when morphine addiction became well understood at the beginning of the 20th century were regulations imposed for its withdrawal as an over-the-counter medicine.

THE PRODUCTION OF MORPHINE IN THE NATURAL WORLD

Morphine is an organic chemical compound found in nature, specifically in the poppy plant (*Papaver somniferum*; Figure 5.1). The plant is typically found growing in arid climates, for example in Afghanistan, within a temperature range of 7–23°C and where the pH of the rich moist soil is 4.5–8.

FIGURE 5.1 Poppy, *Papaver somniferum,* showing the petals and pods.

Morphine (Figure 5.2) is the most abundant alkaloid found in opium. Opium is a complex mixture of chemicals from the poppy plant which contains sugars, proteins, fats, wax, latex, gums, water, ammonia, lactic acid and numerous alkaloids, most notably morphine. Morphine content is generally 8 to 17% of the dry weight of opium, although specially bred cultivars reach 26%. Opium poppy contains at least 50 different alkaloids, but most are in very low concentration. Morphine is the main alkaloid. All alkaloids, including morphine, can be purified initially from crude extracts by the acid-base extraction technique summarized in the chapter on 'Maca from the High Andes in South America' in Part IV and in the chapter on the Madagascan periwinkle entitled 'Africa's Gift to the World' in Part II.

The harvesting of opium, even in modern times, is carried out in the traditional way. The seed case remains on the stem after the flower petals fall and two or three vertical slots are made into the skin using a sharp blade. An incision, if skin deep, allows the latex to ooze out slowly and then to harden on the outside of the pod. The next day the hard latex is scraped off, before new incisions are made to obtain more poppy juice.

PURIFICATION, CHEMICAL COMPOSITION AND PROPERTIES OF MORPHINE

Morphine was first isolated from opium in 1805 by a German pharmacist, Wilhelm Sertürner. He named it morphium—after Morpheus, the Greek god of dreams. The chemical structural formula of morphine was only determined in 1925. There are at least three ways of synthesizing morphine from starting materials such as coal tar and petroleum distillates but the vast majority of morphine is derived from the opium poppy.

Morphine is purified from the opium resin in the following way: first, the resin is soaked in diluted sulfuric acid, which releases the alkaloids into solution without altering the alkaloid molecules themselves. The alkaloids are then precipitated by adding alkaline inorganic compounds—either ammonium hydroxide or sodium carbonate. This last step also separates morphine from other opium alkaloids.

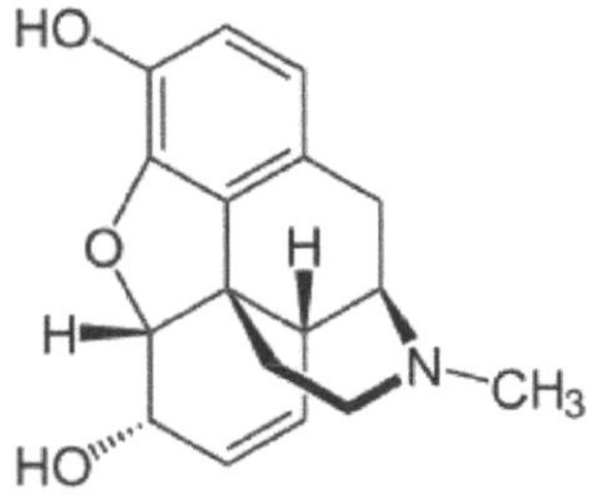

FIGURE 5.2   Chemical structure of morphine.

Chemical Composition and Properties of Codeine

After sufficient quantities of pure morphine have been retained commercially for use as a strong analgesic (pain killer), the remaining morphine is converted into codeine. Codeine does revert to morphine in the human body, producing a milder analgesic effect than pure morphine (220mg of codeine is as powerful as only 30mg of morphine). Whilst codeine remains potentially addictive, if abused, it is much safer to use than morphine. Codeine is by far the most commonly used opioid in the world. As well as its use as an analgesic, codeine is dissolved in a syrupy liquid for use as a cough remedy.

Although codeine is found together with morphine, the concentration of the former in raw opium is much lower. Morphine can be converted to codeine using acetic anhydride, $(CH_3CO)_2O$. It is the hydroxyl group on the benzene ring of morphine that is acetylated.

Acetylation refers to the process of introducing an acetyl group (CH_3CO) into a compound. A reaction involving the replacement of the hydrogen atom of a hydroxyl group of an alcohol with an acetyl group yields an ester, the acetate. Acetic (ethanoyl) anhydride is commonly used as an acctylating agent (see also the chapter on the emergence of aspirin, 'Europe Solves a Headache' in Part II). The following reactions, involving propanol and acetic (ethanoic) acid and propanol and acetic anhydride, would be much more vigorous with acetic anhydride:

$$C_3H_7OH + CH_3COOH = CH_3COOC_3H_7 + H_2O.$$

$$2C_3H_7OH + (CH_3CO)_2 = 2CH_3COOC_3H_7 + H_2O.$$

Morphine can be converted to heroin by this acetylation reaction.

Chemical Composition and Properties of Heroin

A major search by scientists in Germany began in the late 19th century for a powerful alternative to opium and morphine and to find a form of the drug that could be taken orally instead of by injection.

Morphine can be converted to heroin (Figure 5.3). However, heroin is classified as an illegal recreational drug as it is highly addictive. In 1874, when morphine and acetic acid were boiled together, a chemical known as diacetylmorphine or morphine diacetate, also known as heroin, was produced. This is another example of an acetylation reaction. Both of the hydrogen atoms in the hydroxyl groups in morphine are replaced by acetyl groups, CH_3COO, to form heroin.

Heroin is approximately 1.5 to 2 times more potent than morphine weight-for-weight as it is able to cross the blood-brain barrier faster than morphine, subsequently increasing the reinforcing component of addiction. The drug is converted back to morphine in the body before it binds to brain-tissue receptors.

Heinrich Dreser, who was in charge of drug development at Bayer, tested the new semi-synthetic drug on animals and humans—even on himself. He pronounced heroin an effective treatment for a variety of respiratory ailments, especially bronchitis, asthma and tuberculosis. Since it was more potent than morphine, Bayer launched heroin in 1898 under its trademark as a new 'wonder drug— the sedative for coughs' (Figure 5.4). Bayer was soon enthusiastically selling heroin in dozens of countries. Free samples were handed out to physicians. Sadly, the medical profession remained largely unaware of the potential risk of addiction for years. Eventually, doctors began to notice their patients were consuming inordinate quantities of heroin-based cough remedies. It transpired that heroin was not the miracle-cure that some of its early promoters had supposed and Bayer finally halted production in 1913.

Medicinal Uses of Morphine

Morphine is used to relieve severe or agonizing pain and suffering and acts directly on the brain. It appears to mimic endorphins, natural substances produced by the brain that are responsible for reducing pain, which also cause sleepiness and feelings of pleasure in the body. Endorphins are released in response to pain, strenuous exercise or excitement. In clinical settings, morphine exerts its principal pharmacological effect on the brain. Its primary actions of therapeutic value are to reduce pain and make patients sleepy.

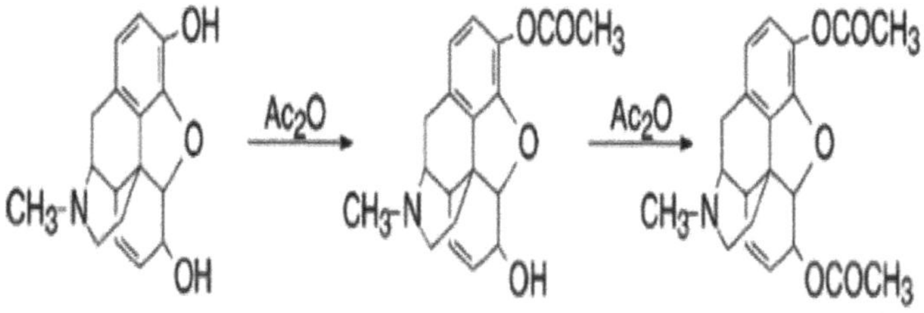

MORPHINE **3-ACETYLMORPHINE** **HEROIN**

FIGURE 5.3 Acetylation of morphine to codeine and then to heroin using acetic anhydride.

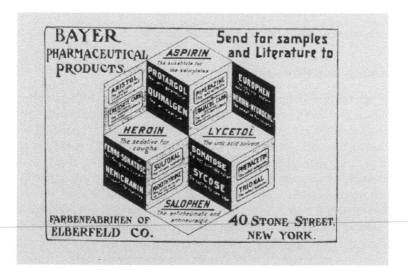

FIGURE 5.4 Bayer's heroin was sold legally at the turn of the 20th century.

Morphine became a controlled substance in the United States under the Harrison Narcotics Tax Act of 1914. Possession without a prescription in the United States became a criminal offence. Morphine was the most commonly abused narcotic analgesic (pain killer) in the world until heroin was synthesized and came into use. Even today, morphine is the most sought-after prescription narcotic by heroin addicts when heroin is scarce.

SUMMARY

Morphine is one of nature's great botanical miracles. It has powerful pain killing effects. When used appropriately under medical supervision, it has been a life saver and has given relief to millions of people throughout the world. Furthermore, the discovery of codeine as an anti-tussive compound (cough suppressant) has been a blessing to many people. Even though there are side effects, it appears the benefits outweigh the risks. However, this cannot be said of heroin use, which has been a curse on societies and a destroyer of lives.

In the world of good drugs, bad drugs, this remains a tale of two parts:

- most morphine is derived from the opium poppy
- morphine, opium, codeine and heroin are related chemicals known as alkaloids
- morphine, opium and heroin are considered narcotic drugs and are controlled substances in most countries around the world
- heroin use is illegal
- morphine is a pain killer
- codeine alleviates or suppresses coughing
- codeine is made from morphine by chemical conversion.

Questions

1. Alkaloids are produced in nature by a large variety of organisms including bacteria, fungi, plants and animals. Why is this so?
2. Where are the major poppy growing countries in the world?
3. Why does the drug industry seek orally acting drugs?
4. How does the structure of codeine differ from that of heroin?
5. Look up the properties of the organic compound thebain. How is it different from morphine? Why is thebain an important alkaloid although not as biologically active as morphine?

SUGGESTED FURTHER READING

J. Mann. 2009. *Turn On and Tune In: Psychedelics, Narcotics and Euphoriants*. RSC Publishing.
C. Trocki. 1999. *Opium, Empire and Global Political Economy—A Study of the Asian Opium Trade 1750–1950*. Routledge; Taylor and Francis.

CANNABIS AND MARIJUANA

Abstract: *An inebriant and a modern social and medical enigma, cannabis—source of marijuana and hashish—is among the oldest of cultivated plants dating back to the beginnings of agriculture. Praised and maligned throughout history, cannabis continues to beguile with its subtle magic of mystery and medicine.*

Organic Chemistry

- *isomers*
- *elimination reaction*
- *condensation reaction*
- *mono-terpenes and the isoprene rule*
- *cyclic terpenes.*

Context

- *cannabinoids*
- *terpenes.*

CANNABIS

The use of fibrous material from cannabis goes back at least 10,000 years and the consumption of cannabis as an intoxicant for at least 5,000 years. Charred seeds have been found in an ancient ritual brazier in Romania while cannabis leaves and seeds have been discovered buried with a Chinese mummy.

Cannabis is a member of the hemp family of plants. Its fibers were turned into paper in China while the leaves were dried to form 'grass', which was then smoked for relaxation or used in teas for medicinal purposes.

In India, where the cannabis plant is known as *ganja*, a preparation of marijuana called *bhang* was made from the leaves and drunk with milk or water for its hallucinogenic effects. Extracts from the flowering heads were used in the treatment of anxiety as early as 1,400 BC.

Although the use of marijuana is controversial today, it does have some beneficial properties that have led to legalization for medicinal purposes in many countries. Specifically, cannabis is an established treatment for encouraging appetite in cancer patients. It is also utilized to help with nausea, weight-gain, neuralgic pain and glaucoma.

INTRODUCING CANNABINOIDS

The cannabis genus has two economically important species: *Cannabis sativa* and *Cannabis indica*. These plants yield a sticky resin known as *hashish*, which contains a class of diverse compounds, some possessing hallucinogenic properties, known to science as cannabinoids. One particular cannabinoid is δ 1-tetrahydrocannabinol (THC; Figure 5.5).

THC is the primary psycho-active compound in marijuana and forms the natural biochemical defense of the cannabis plant against herbivores and disease. The discovery of THC was made by a team of researchers from the Hebrew University Pharmacy School, Israel in 1964. Nowadays, synthetic THC is manufactured in California. It has been approved for limited medical use by the Federal Drugs Administration in the USA. However, over 60 naturally occurring cannabinoids have been identified to date in cannabis. Several of these cannabinoids—in addition to naturally occurring terpenoids (oils) and flavonoids (phenols)—have also demonstrated therapeutic qualities. In fact, experiments have shown that the full range of psychoactive and medical effects of cannabis resin cannot be re-created by the singular use of pure synthetic cannabinoid drugs such as THC. The indications are that other components of cannabis resin such as terpenes are either psychoactive themselves or are able to modulate the effect of cannabinoids when they are ingested together. Cannabinoids, such as THC, are chemically classified as terpenoids.

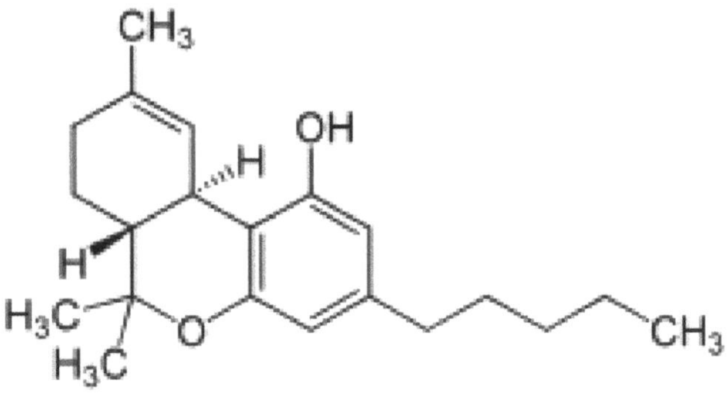

FIGURE 5.5 δ 1-tetrahydrocannabinol (THC) is the main bioactive constituent of cannabis.

TERPENES

Terpenes are a huge and varied class of hydrocarbons that make up a majority of plant resins and saps. Essential oils (see the chapter on 'European Lavender' in Part VI), composed primarily of terpenes, have a long history of medicinal use. A familiar terpene containing 40 carbons is β-carotene (see the chapter on 'Saffron and Carotenoids' in Part VII), which is responsible for the orange color of carrots and is a source of vitamin A. The low molecular weight terpenes, however, are very volatile and many can be recognized from their distinctive smell. Terpenes give pine trees and lemons their scent (pinene and limonene). Apart from use in fragrances (see the chapter on 'European lavender' in Part VI), terpenes may be employed as an organic solvent. Oil of turpentine, a mixture of terpenes derived from pine trees, is an example.

Terpenes could find application as a replacement for crude oil as they are purely hydrocarbons. Many of the materials used today, whether as fuels, plastics or deodorants, are derived from chemicals found in oil. Research is taking place to try to discover alternative, sustainable sources of basic substrates that can reduce or remove our dependency on oil. Substances in plants can be extracted and transformed into these industrial substrates. Sometimes these substrates are referred to as platform chemicals. They are simple and cheap and are used by the industry as starting materials from which more complex and valuable chemicals may be made.

CLASSIFICATION OF TERPENES

Terpenes are polymers and are found mostly in plants (see also the chapters on 'Saving the Pacific Yew' in Part II and the chapters on 'Frankincense and Myrrh' and 'Camphor' in Part VI) although larger and more complex terpenes (e.g. squalene) do occur also in animals as steroids.

Terpenes are made up of multiples of isoprene molecules. Isoprene is an alkene having the molecular formula, $CH_2.C (CH_3).CH.CH_2$, which has two carbon–carbon double bonds within a short five-carbon chain.

The simplest of all terpenes consist of two isoprene units linked together and has the molecular formula $C_{10}H_{16}$.

Terpenes are classified by use of an empirical feature known as the isoprene rule. Molecules of terpenes, $(C_5H_8)n$, such as squalene, presented later, are made up from linear multiples (n) of smaller isoprene molecules, (C_5H_8), which each contain five carbon atoms (where n is an integer).

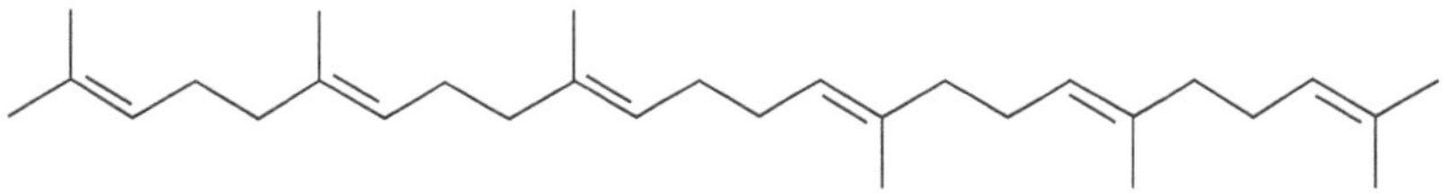

If $n = 2$, the molecule is described as a monoterpene.
If $n = 3$, the molecule is described as a sesquiterpene.
If $n = 4$, the molecule is described as a diterpene, etc.

Terpenes can undergo natural biochemical modification through oxidation or complex rearrangement reactions to produce a variety of open chain and cyclic terpene compounds (see the chapters on 'Frankincense and Myrrh' and 'Camphor' in Part VI). In fact, a wide variety of cyclic terpenes is known. As terpenes, these molecules also consist of multiples of the building block C_5H_8. However, cyclic terpenes do have fewer carbon–carbon double bonds than open chain or acyclic terpenes. An example of a cyclic monoterpene is the fragrant substance limonene, (Figure 5.6), which may be obtained from the rind of lemons.

TERPENES AND ELIMINATION REACTIONS

In an elimination reaction, a small group of atoms or a small molecule break away from a larger molecule and are not replaced.

Dehydration is a common elimination reaction in nature. Elimination reactions, catalyzed by enzymes, lead to the formation of terpenes from carbohydrates and fatty acids. Although the details of the mechanisms of many of these reactions are convoluted and little understood, the overall effect is the elimination of water, hence the oxygen, from the carbohydrate or fatty acid. An illustration of dehydration in a simple elimination reaction, which can be performed in the laboratory, is the formation of the alkene 2 butene, when the corresponding alcohol, 2 butanol, is heated in the presence of a strong acid such as sulfuric acid.

$$CH_3.CHOH.CH_2.CH_3 = CH_3.CH.CH.CH_3 + H_2O$$

In the laboratory, elimination reactions are an effective means of introducing the functional group of the alkenes into molecules. Since alkenes are so reactive, they form a good platform from which to make many other organic chemicals. The elimination reaction is, therefore, a good starting point on a pathway toward the production of a whole array of organic compounds.

THE CONDENSATION REACTION

An elimination reaction is similar to a condensation reaction.

In a condensation reaction, the molecules of two compounds combine together with the loss of a molecule of water—that is why it is called a condensation reaction. There are many examples of condensation reactions in organic chemistry.

Condensation polymerization usually involves two different types of monomer. Each monomer has at least two functional groups—often at either end of the monomer. Each functional group reacts

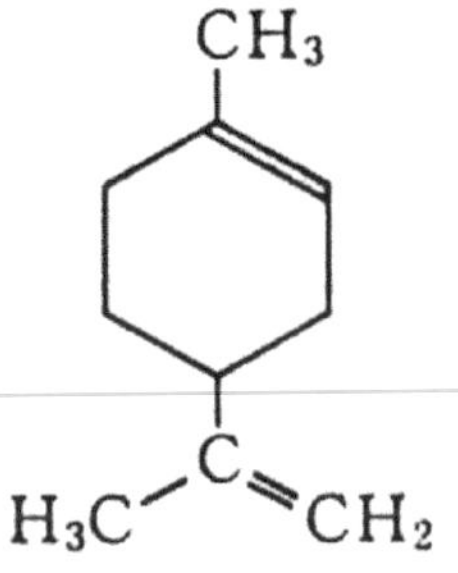

FIGURE 5.6 Limonene.

with a different functional group on a neighboring monomer to form a link that builds successively into a polymer chain. Examples of synthetic polymers formed from condensation reactions are polyesters and polyamides used in modern textiles and, of course, the polypeptides and proteins formed in the natural world presented in the chapter on 'Wheat—Ancient and Modern' in Part III.

Questions

1. Legalization of marijuana? For and against—discuss!
2. An illustration of dehydration is the formation of the alkene 2 butene, when the corresponding alcohol, 2 butanol, is heated in the presence of a strong acid such as sulfuric acid.
    $CH_3.CHOH.CH_2.CH_3 = CH_3.CH.CH.CH_3 + H_2O$
    Give examples of other elimination reactions in organic chemistry that involve the loss of small molecules other than water.
3. A polyester may be formed from reaction between the molecules of a dicarboxylic acid, COOH.R.COOH and the molecules of a diol, OH.R*.OH, where R and R* represent hydrocarbon chains.
    Draw the repeating unit of the polyester.
    Explain why this process is an example of a condensation reaction.
    Now, choose an amine and draw the repeating unit of the associated polypeptide.
4. The simplest of all terpenes consist of two isoprene units linked together and has the molecular formula $C_{10}H_{16}$. Provide a molecular structure for an isomer of this terpene.
    Explain the term addition polymerization.

SUGGESTED FURTHER READING

P. M. Richardson. 1986. *Flowering Plants—Magic in Bloom. Encyclopaedia of Psychoactive Drugs*. Chelsea House Publishers.

COCA AND COCAINE

Abstract: Over many centuries, coca was both a sacred herb and a staple part of diet for the indigenous peoples of the Andes (Colombia, Ecuador, Peru, and Bolivia). Coca is the source of the useful but abused alkaloid, cocaine, which was first sold legally in the 19th century only to become listed as an illicit narcotic drug in the following century. And yes, with cocaine removed, coca is still a flavoring ingredient in Coca Cola®!

Organic Chemistry

- *alkaline nature of alkaloids*
- *salt formation and acid-base extraction.*

Context

- *the alkaloid, cocaine*
- *organic compounds containing nitrogen in a carbon ring.*

COCA AND THE COCA PLANT

The coca bush has been a domesticated plant since early times in South America. Little was known of the narcotic properties of the leaf outside the Andes. People indigenous to South American chewed the leaves of the coca plant, *Erythroxylon coca*, as they are a source of vital nutrients in addition to numerous alkaloids, among them cocaine. Remains of coca leaves have been found interred with ancient Peruvian mummies, presumably because of belief in the value of coca in the after-life. Chewing coca leaf was—and remains to the present day—a widespread practice in many indigenous communities in the high Andes. Although the stimulant and hunger-suppressant properties of coca had been known for many centuries, it was not until 1855 that the alkaloid responsible, cocaine, was first isolated (Figure 5.7). The first synthesis of cocaine was completed soon afterwards, followed by elucidation of its structure in 1898.

COCAINE

Cocaine acts as a powerful stimulant lifting mood sharply, but it is quite addictive, leading to a craving for more. A feeling of well-being, even euphoria, is quickly followed by contrasting emotions—edginess, anxiety, paranoia and depression. Over time, abusers of cocaine experience a wide and disparate range of quite noxious physical effects that are too numerous to mention individually but do include conditions associated with the cardiovascular toxicity of the drug, which involve heightened risk of cardiac arrest.

Cocaine was first traded as a legal drug in the 19th century, but as its damaging properties became better understood it was proscribed as an illicit narcotic in the 1980s. Now, production, distribution and sale of cocaine and cocaine products are restricted and illegal in most countries.

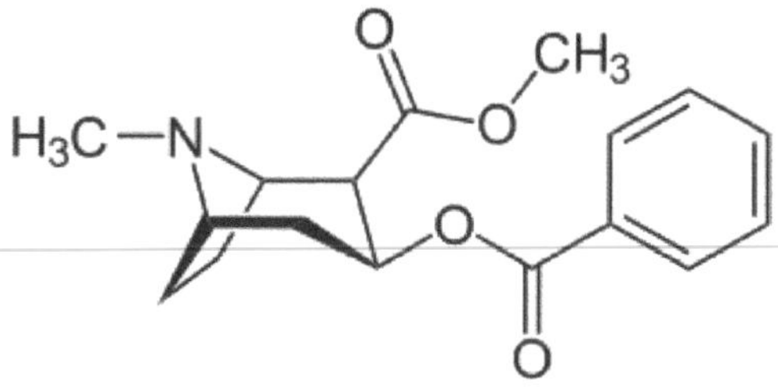

FIGURE 5.7   The alkaloid cocaine.

The drug is regulated by the United Nations Convention against Illicit Traffic in Narcotic Drugs and Psychotropic Substances. Additionally, many countries have passed their own legislation. In the United States, the manufacture, importation, possession and distribution of cocaine is regulated by the 1970 Controlled Substances Act. As a consequence of suppression, it is organized criminal cartels that dominate the supply of cocaine. Coca is grown and processed into cocaine in South America (particularly in Colombia, Bolivia and Peru) and, because of its high black-market price, is smuggled worldwide, particularly into developed countries such as those in Europe and North America.

A synthetic means of producing cocaine would be highly desirable to illegal drug traffickers for obvious reasons, as it would eliminate their dependence on unreliable offshore sources and international smuggling. Fortunately, synthesis of economically significant quantities of cocaine in the laboratory is very difficult. The complex structure of the molecule gives rise to the unavoidable formation of many different enantiomers that are physiologically inactive, thereby severely limiting both the yield and purity of the product.

CHEMICAL PROPERTIES OF COCAINE

Alkaloids are a group of naturally occurring chemical compounds that contain nitrogen atoms within cyclic rings of carbon and hydrogen atoms. Alkaloids may also contain oxygen and sulfur. The nitrogen atoms in alkaloids behave in a similar fashion to amines (see the chapters on 'Coffee—Wake-up and Smell the Aroma' in Part II and on 'Tobacco' in Part V) in that they are basic, although only weakly so. Curiously, most alkaloids have a bitter taste.

Although alkaloids are produced by a large variety of organisms many of them are toxic to other organisms. In humans, many alkaloids exhibit pharmacological effects and have been used for centuries as medication, as recreational drugs and in tribal rituals. Various alkaloids act on a diversity of metabolic systems in humans and several examples are covered in this book: the local anesthetic and stimulant cocaine; the stimulants caffeine and nicotine; the analgesic morphine; the anti-cancer compound vincristine and the anti-malarial drug quinine.

Cocaine is an alkaloid derived from a natural building block called tropane. The alkaloids of tropane occur in certain families of plants, notably in the *Erythroxylaceae*, which includes coca and also in the *Solanaceae*, which includes henbane, deadly nightshade, potato and tomato.

Tropane (Figure 5.8) is a bi-cyclic hydrocarbon with a single nitrogen atom bridging both a five membered and a six membered ring. In this respect, tropane is very similar to another heterocyclic, saturated substance, which contains a nitrogen atom in the ring, namely piperidine, $C_6H_{10}N$, which is obtained from an extract of black pepper and is a secondary amine. Piperidine is used as a chemical building block for synthetic manufacture in the pharmaceutical industry.

As an alkaloid and a cyclic secondary amine, cocaine is a weakly basic compound, which can combine with stronger acids, such as inorganic acids, to form salts: the hydrochloride sulfate and

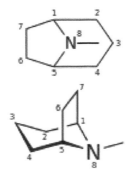

FIGURE 5.8   Molecular structure of tropane.

nitrate. The salts of cocaine are polar compounds. Therefore, they dissolve readily in a polar solvent such as water. In marked contrast, molecules of pure cocaine are covalent and are practically insoluble in water yet are readily soluble in organic solvents.

ISOLATION OF COCAINE

Acid-base extraction is a procedure using sequential liquid-liquid extractions to purify acids or bases from mixtures by exploiting their chemical properties. The procedure works only for acids or bases with a large difference in aqueous solubility between their charged and their uncharged forms. Acid-base extraction can be performed as a routine process in the laboratory in order to isolate natural products such as specific alkaloids from crude extracts since alkaloids are weakly basic substances. Hence, the technique may be applied to isolate samples of pure cocaine. This procedure is described extensively in the chapter entitled 'Africa's Gift to the World' in Part II and in the chapter on 'Maca from the High Andes in South America' in Part IV.

LEGITIMATE APPLICATIONS OF COCAINE

It should be noted that with the approval of the American College of Medical Toxicology, some limited and controlled medical applications of cocaine are permitted. It is used by some physicians to staunch strong nosebleeds in patients and as an anesthetic before minor nasal surgery. Again, due to its property as an anesthetic, dental surgeons may use it before oral procedures.

Questions

1. In looking at the structure of a molecule of cocaine, identify the functional groups and building blocks present.
2. Explain some of the chemical properties of cocaine by reference to the chemistry of a saturated heterocyclic compound which contains nitrogen in the ring, namely piperidine.
3. Why is acid-base extraction helpful in purifying alkaloids?

SUGGESTED FURTHER READING

H. Hobhouse. 2005. *Seeds of Change. Six Plants That Transformed Mankind*, pp. 291–363. Counterpoint.
J. Mann, J. Emsley, P. Ball, P. Page, J. P. Michael, and H. Oakeley. 2009. *Turn on and Tune In: Psychodelics, Narcotics and Euphoriants*. Royal Society of Chemistry.
D. F. Rhoades. 1979. *Evolution of Plant Chemical Defense against Herbivores in Herbivores: Their Interaction with Secondary Plant Metabolites*. Academic Press.

TOBACCO: A PROFOUND IMPACT ON THE WORLD

Abstract: *Tobacco, once introduced to the Old World, changed the habits of mankind forever, creating a powerful smoking habit through cigarettes, pipes and cigars, required slavery to meet ever growing demand and, as modern science can attest, has led to nicotine addiction and to increased incidence of lung disease.*

Organic Chemistry

- *primary, secondary and tertiary amines*
- *heterocyclic aromatic substances containing nitrogen or sulfur in the ring*
- *amides.*

Context

- *nicotine*
- *pyridine and pyrrole as chemical building blocks.*

TOBACCO

Tobacco is an agricultural product processed from the leaves of plants in the genus *Nicotiana* of the *Solanaceae* family. The product manufactured from the leaf is used in cigars, cigarettes, snuff, pipe and chewing tobacco. The chief commercial species, *Nicotania tabacum*, is believed to be native to tropical America. *Nicotinia rustica*, which is a mild-flavored and fast-burning species, was the tobacco originally grown in Virginia, although it is now cultivated chiefly in Turkey, India and Russia. The alkaloid nicotine (Figure 5.9) is the most characteristic constituent of tobacco and is responsible for its addictive nature. The usage of tobacco today is practiced by possibly up to one third of the adult population worldwide.

The tobacco plant grows to a height of five feet and is harvested once a year. The leaves are cut down and have medicinal value as an analgesic applied to wounds and to snake and scorpion bites. The leaves were collected by natives of the New World—especially those of *Nicotinia tabacum*. The tobacco strain, *Nicotinia rustica*, spread through Central America northwards to North America and is believed to be the tobacco known in and around the Mississippi valley during the 1st century BC. Scientific evidence points to the fact that smoking and chewing tobacco was first practiced on the American continent.

HISTORICAL NOTE

The Mayan Empire existed from the 3rd to the 9th centuries AD. The Mayans lived in the Central American region, which we now know recognize as the Yucatan peninsula of Mexico, Guatemala, El Salvador and Honduras. When the Spanish arrived they found the Indian people to be ardent smokers using an herb unknown to the Western world. It was smoked recreationally as a pastime and also had significant religious and mythological implications. At the time of the conquest there was no cultivation but today it is most certainly a very important cash crop.

Columbus noted in his journal of 1492 that the San Salvadorian natives brought fruit, wooden spears and dried leaves (tobacco) with a distinctive fragrance. The local people smoked dried rolled leaves. The Spanish and later new arrivals to the New World took up the smoking habit, which soon spread quickly among them.

The Spanish brought tobacco back to Europe although initially there was resistance to its use by the clergy. Nevertheless, by 1560, tobacco was being exported to Europe by Portuguese, Spanish, Dutch and English traders. In 1594, the historian van Meteren noted the use of tobacco in Holland: 'from diverse nations West Indies, Brazil and Peru, a dried herb called nicotania leaf (in the Indies known as tobacco) which was smoke-dried and used in a pipe lit with a live coal or candle'.

Historically, the indigenous people were smoking a cigar form with large, rolled tubes of dried leaf. It appears that as tobacco use went northwards the preferred form of smoking was by pipe. The indigenous people of Mexico today are still predominantly cigar and cigarette smokers.

While tobacco had long been known in the Americas, it was not until the arrival of Europeans in North America that tobacco became a widely abused drug and a very important item of trade. Its popularity initially stimulated the development of the economy of the southern states of the United States until it was superseded by cotton as an important cash crop. Following the American Civil War, changes in consumer demand and in the structure of the labor force allowed for the development of the cigarette and quickly led to the growth of tobacco companies.

Tobacco influenced profoundly both the New World and the Old World through commerce, simultaneously creating wealth for some and impacting adversely upon the health of others. The worldwide habit of smoking tobacco is a remarkable phenomenon in the cultural history of mankind.

NICOTINE

Nicotine (Figure 5.9) was first isolated in Germany from the tobacco plant in 1828 by the physician Wilhelm Heinrich Posselt and the chemist Karl Ludwig Reimann. Its molecular structure was elucidated in 1893 by Adolf Pinner and Richard Wolfenstein. The compound contains two nitrogen atoms, one present in a larger pyridine ring and the second in a pyrrole ring. The systematic name for nicotine is 3-(1 methyl-2 pyrroldinyl) pyridine.

Nicotine is a hygroscopic, colorless, oily liquid that is readily soluble in organic solvents such as alcohol and ether. It is miscible (see Glossary) with water between 60°C and 210°C. As a heterocyclic aromatic amine, nicotine forms salts and double salts with acids that are solid at room temperature and are water soluble.

Nicotine (Figure 5.9) is an alkaloid and contains a nitrogen atom in a saturated ring of carbon atoms. It is a tertiary amine. Thus, a short section on amines in general is warranted and discussed in detail later.

AMINES

Amines are ubiquitous in the natural world. Many important molecules are based on amines such as amino acids. Amines are utilized industrially as building blocks in the manufacture of dyes and pharmaceutical products.

Amines are compounds characterized by the presence of the nitrogen atom, a lone pair of electrons and three substituents. Amines are derivatives of ammonia in which one or more of the hydrogen atoms have been replaced with substitutes. Due to the lone pair of electrons, amines are basis substances. The degree of basicity can be influenced by neighboring atoms, stereochemistry and the degree of solubility in water of the cation produced.

Amine molecules can form hydrogen bonds. They are, therefore, soluble in water and have elevated melting and boiling points in relation to their mass.

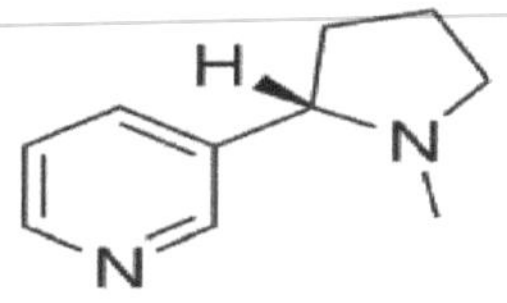

FIGURE 5.9 Nicotine, an alkaloid found in tobacco.

Amines are classified according to the replacement of the hydrogen atoms bonded to the nitrogen atom in the following order:

- one substituent—primary amine
- two substituents—secondary amine
- three substituents—tertiary amine.

Amines are prepared as building blocks as the basis of a variety of industrial processes by alkylation with alcohols or by the reduction of nitriles by hydrogen.

In the case of cyclic amines, the nitrogen atom has been incorporated into a ring of carbon atoms effectively making the amine either like a secondary amine (in an aromatic ring) or a tertiary amine (in a non-aromatic ring).

Unsurprisingly, an amine group bonded to an aromatic group is known as an aromatic amine. The two functional groups influence one another: the aromatic structure decreases the alkalinity of the amine by the delocalization of electrons while the amine group significantly increases the reactivity of the ring toward electrophilic agents for the same reason.

HETEROCYCLIC AROMATIC COMPOUNDS

Heterocyclic compounds are exceedingly numerous. In fact, their chemistry forms one of the largest subdivisions of organic chemistry.

Pyridine

One of the most frequently occurring heterocyclic aromatic compounds is pyridine (Figure 5.10), which has a structure very similar to that of benzene. However, the six-member ring contains a tri-valent nitrogen atom.

Pyridine is a colorless liquid with an unpleasant, characteristic odor. It is completely miscible with water and most organic solvents, and is itself a very good solvent. As it is a stable, aromatic amine, pyridine is a moderate base in that it forms water-soluble salts, such as pyridine hydrochloride in the presence of hydrochloric acid.

Pyrrole

Pyrrole (Figure 5.11) is a colorless, volatile liquid that darkens on exposure to oxygen in the air. Pyrrole, C_4H_5N, is aromatic and has a five membered ring.

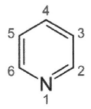

FIGURE 5.10 Molecular structure of pyridine.

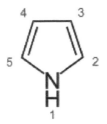

FIGURE 5.11 Molecular structure of pyrrole.

As an aromatic amine, pyrrole is not very nucleophilic, being only weakly basic at the nitrogen atom. The pyrrole skeleton occurs commonly in molecules in the natural environment, for example, chlorophyll and vitamin B12 — (see the chapter on 'Our World of Green Plants—Human Survival' in Part VII for more on chlorophyll).

Both pyridine and pyrrole are used extensively as chemical building blocks in the organic chemicals and pharmaceutical industries.

Thiophen

Another common heterocyclic compound is thiophen (Figure 5.12), which possesses a five-member ring containing one divalent sulfur atom (see the chapter on 'Garlic and Pungent Smells' in Part IX for more on organic compounds that contain sulfur).

Owing to the delocalization of electrons over the ring structure, thiophen is remarkably stable. At this point, it will be worth referring to the properties of benzene described in the chapter entitled 'Central America's Humble Potato' in Part II. Pyridine, pyrrole and thiophen all show chemical behavior similar to that of benzene and can undergo substitution reactions when hydrogen atoms on the ring are replaced by other atoms or groups.

Amides

An important chemical reaction of amines is the formation of amides with ketones or aldehydes or more readily with acyl chlorides. The resulting compounds have a molecular structure in which a nitrogen atom is attached to a carbonyl group. These compounds are known as amides and have the general formula $R_1.CO.NH_2$. In the case of amides substituted at the nitrogen atom, namely N-substituted amides, the general formula is $R_1.CO.NR_2R_3$.

Due to the higher electro-negativity of the oxygen atom in comparison with that of the nitrogen atom, the carbonyl group is slightly dipolar with the carbon atom slightly positive, as electron density is displaced to the oxygen end of the covalent bond. As a consequence, carbonyl bonds react at the carbon atom with electron-rich entities, such as the nitrogen atom in the molecules of ammonia and amines, to produce amides.

Amines can be readily acylated by acyl chlorides to form amides and N-substituted amides.

$$CH_3.CO.Cl + C_3H_7.NH_2 = CH_3.CO.NH.C_3H_7 + HCl$$

ethanoyl chloride + propylamine = N-propylethanamide + hydrochloric acid

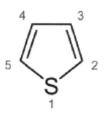

FIGURE 5.12   The molecular structure of thiophen.

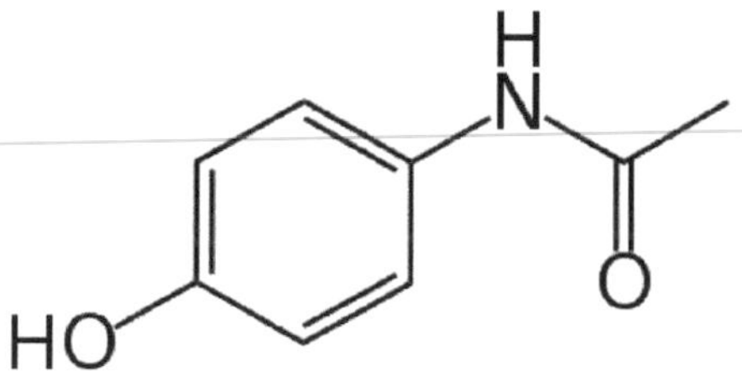

FIGURE 5.13   Acetaminophen or paracetamol.

$$CH_3.CO.Cl + C_3.H_7.NH_2 = CH_3.CO.NH.C_3.H_7 + HCl$$

ethanoyl chloride + propylamine = N-propylethanamide + hydrochloric acid

$$CH_3.CO.Cl + C_6.H_5.NH.CH_2 = CH_3.CO.NH.C_6.H_5 + HCl$$

ethanoyl chloride + phenylamine = N-phenylethanamide + hydrochloric acid

Amides are common in the natural world. An example of a substance with an amide group is urea, NH_2CONH_2, which is one of the compounds excreted in urine as a result of the metabolic breakdown of proteins. Urea is used as an organic fertilizer in farming and historically in the tanning of animal hides to make leather. Amides are a major component of proteins and enzymes too—see the chapter entitled 'Wheat—Ancient and Modern' in Part III.

An example of a synthetic amide of significant medical value for mild pain relief and fever reduction is known commercially as acetaminophen in the USA (but as paracetamol elsewhere) with a structure $CH_3.CO.NH.C_6H_5OH$, which can be formed from a condensation reaction involving para-aminophenol and the acyl chloride, ethanoyl chloride.

PROPERTIES OF NICOTINE

Nicotine is a hygroscopic, colorless oily liquid that is readily soluble in organic solvents such as alcohol and ether. It is miscible (see Glossary) with water between 60°C and 210°C. As a heterocyclic aromatic amine, nicotine forms salts and double salts with acids that are solid at room temperature and are water soluble.

The fused pyridine and pyrrole rings in nicotine are aromatic with electrons delocalized over the carbon-nitrogen skeleton, which helps to explain its weakly basic character and solubility in water.

Nicotine is optically active having two enantiomeric forms. The naturally occurring form of nicotine is levorotatory ((−)-nicotine) while the synthesized form, first prepared in 1904, is dextrorotatory ((+)-nicotine) and is less active physiologically. Evidently, stereochemistry plays an important part in the activity of (−)-nicotine.

On exposure to ultraviolet light or various oxidizing agents, nicotine is converted to nicotinic acid (vitamin B3) amongst other products.

USES OF NICOTINE

Agriculture

Tobacco plants produce nicotine as a natural insecticide and this can be concentrated for use as an artificial insecticide in order to improve the yield of crops grown in a monoculture. Neo-nicatinoids are synthetic analogues of the natural insecticide nicotine. They are broad-spectrum systemic insecticides, which are applied as sprays to crops or as seed and soil treatments.

In contrast to agricultural methods supporting the growth of an economically important crop in a monoculture, organic farming relies on techniques such as crop rotation, use of compost for soil fertility and biological pest control. In the USA, the National Organic Standards Board (NOSB) defines the practice as follows:

> Organic agriculture is an ecological production management system that promotes and enhances biodiversity, biological cycles and soil biological activity. It is based on minimal use of off-farm inputs and uses management practices that restore, maintain and enhance ecological harmony.

Yields of crop per acre are lower but organic farming is a sustainable practice. In recent decades, despite economic factors, the market for organic food and related products has grown rapidly. Demand has driven a corresponding increase in organically managed farmland which has grown steadily in extent though still representing a relatively small proportion of farmland in cultivation worldwide.

Human Health and Pharmacy

There appears to be little compelling evidence that use of nicotine is related to a substantially increased risk of cancer (Whelan, 2014). Nicotine is a stimulant producing a short-term increase in blood pressure and pulse rate, which could affect general health. Although nicotine is relatively safe for most individuals, it may have a negative effect on fetal development and should be avoided during pregnancy.

It is true that a drop of pure nicotine can be deadly. The nicotine extracted from a pack of cigarettes and concentrated in one dose would be likely to be fatal if it were all ingested at once. Smokers are never exposed to pure nicotine and do not absorb that much. However, the smoke from a burning cigarette contains compounds that have a devastating range of harmful health effects. The damage and health risks come from the smoke that is inhaled, rather than from the nicotine itself. This key fact is highlighted in an article in the American Council on Science and Health entitled 'The Effects of Nicotine on Human Health'. Smokers smoke for the nicotine 'rush', yet risk illness and early death from the effects of tobacco smoke. Ironically, a primary and current therapeutic use of nicotine arises in the treatment of nicotine dependence, in order to eliminate smoking, given the established damage that the practice does to both the health of the inhaler and to that of secondary inhalers. Controlled levels of nicotine are administered to patients through dermal patches, lozenges, electronic substitute cigarettes or nasal sprays in an effort to wean them gradually from dependence.

Nicotine has strong mood-altering effects and can act on the brain as both a stimulant and a relaxant. Once in the bloodstream, nicotine will circulate around the body and reaches the brain within about ten seconds. As is widely appreciated, nicotine is addictive. In the brain, nicotine stimulates the release of neurotransmitters, which relieve negative feelings such as pain and anxiety while pleasant sensations are enhanced. Nicotine is not intoxicating; neither does it appear to impair judgment, motor skills or sociability. Nicotine intake reduces appetite, which over time causes weight loss. Because of these properties, nicotine is being considered as a therapeutic agent in small doses to treat such conditions as attention deficit disorder, Alzheimer's disease, Parkinson's disease, obesity, ulcerative colitis and inflammatory skin disorders.

Questions

1. Explain the nucleophilic substitution reactions of different aromatic heterocyclic compounds such as pyridine and pyrrole in terms of resonant structures and Huckel theory.
2. Draw structural formulae for each of the following molecules:
   * ethylisopropylamine
   * tert-butylamine
   * 2-aminopentane
   * 1,6-diaminohexane and
   * N,3-diethylaniline.
3. Draw diagrams showing how the molecules of anhydrous propylamine form hydrogen bonds to one another in the liquid phase. Then show schematically how the molecules in an aqueous solution of ethylamine would behave.
4. Highlight the methods and economics of organic farming and compare and contrast them with those of standard agricultural practice, which often occurs in monoculture.
5. Name and draw the structure of the molecule that has an oxygen atom instead of a nitrogen or sulfur atom within a five-membered ring.

SUGGESTED FURTHER READING

F. Robisek. 1978. *The Smoking Gods; Tobacco in Maya Art, History and Religion.* University of Oklahoma Press.

REFERENCE

E. Whelan. 2014. *The Effects of Nicotine on Human Health.* Am Coun Sci Heal 2.

Part VI

Exotic Potions, Lotions and Oils

INTRODUCTION

It should come as no surprise that from the earliest times mankind sought chemical extracts from the huge natural resource of plants. By trial and error, discoveries were made concerning potions, lotions and oils, many with cosmetic properties, which make daily life more bearable and enjoyable.

SUGGESTED FURTHER READING

K. Hughes. 2007. *The Incense Bible*. Haworth Press.
A summary of the chemistry in Part VI follows.

Chapter	Organic Chemistry	Context
Camphor	Polymerization	Terpenes—linear and cyclic
	Nucleophilic addition	
	Carbonyl functional group	
	Infra-red spectroscopy	
	Greenhouse gases	
	Global warming	
Frankincense and	Steric hindrance	Cyclic terpenoids
Myrrh	Essential oils	Nomenclature, cyclic terpenes
Lavender	Fractional distillation	Hydrosols and colloids
	Steam distillation	Vegetable oils and fats
	Glycerol	Soaps, Esters, Detergents
	Fatty acids	
	Hydrolysis reactions	
Aloe	Chromatography—	Cosmetic oils
	Paper, Thin Layer, Gas,	Polysaccharides
	Liquid, Ion Exchange	Glycosides, Anthraquinones
	Sustainability and 'fracking'	Galactomannan

DOI: 10.1201/9781032664927-6

A PLANT FROM THE EAST INDIES, CAMPHOR

Abstract: *Camphor is a volatile, waxy, white-to-translucent substance with a strong aromatic smell, which is extracted from the wood of the camphor laurel, Cinnamomum camphora, a large evergreen tree of Sumatra and Borneo.*

Organic Chemistry

- *nucleophiles and nucleophilic addition reactions*
- *the carbonyl functional group*
- *steam distillation.*
- *infra-red spectroscopy and molecular structure determination*
- *sustainability; the'greenhouse' effect, climate change and global warming.*

Context

- *terpenes*
- *cyclic terpenoids.*

THE CLASSIFICATION AND STRUCTURE OF CAMPHOR

Terpenes are a class of molecules that typically contain from ten up to forty carbon atoms, built from a five-carbon building block called isoprene (see also the chapter on 'Saving the Pacific Yew' in Part II and the chapter on 'Cannabis and marijuana' in Part IV).

Camphor (Figure 6.1) is related to terpene and is classified as a terpenoid with the chemical formula $C_{10}H_{16}O$. Its structure reveals the features of a cyclic alkane and a carbonyl functional group.

Terpenoids typically are found as components of plants and are regarded as essential oils, in that they have properties useful to man. Many terpenoids, e.g. menthol and camphor, have medicinal value. A throat lozenge containing menthol will help clear blocked sinuses caused by the common cold. Wood varnish contains a terpene called pinene, which is abundant in pine trees. When pinene is exposed to air and sunlight, it oxidizes and polymerizes slowly to a mixture of terpenoids, which produces a fine hard finish for wood products.

DIVERSE USES OF CAMPHOR

While camphor is poisonous if ingested in large doses, when utilized carefully in small quantities, it has many applications.

The Food and Drug Administration in the USA (FDA) sets a low upper limit for the amount of camphor in consumer products. Medicinal use of camphor is discouraged by the FDA except for skin-related applications, which contain only small amounts of camphor.

In medieval Europe, camphor was used as an ingredient in sweets. In Asian countries today, camphor is still used as a flavoring in confectionery and dessert dishes.

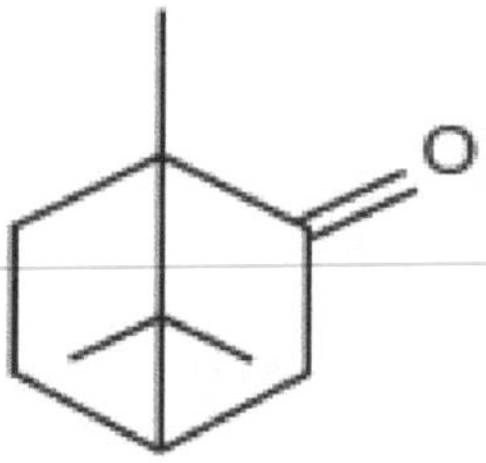

FIGURE 6.1   Structure of camphor.

Camphor is readily absorbed through the skin, produces a soothing sensation of cooling, acts as a gentle local anesthetic and is mildly antiseptic in nature. It is this combination of properties that make camphor an effective ingredient in anti-irritant and cooling gels for the skin. Camphor vapor is also an active ingredient in products, which can suppress coughing by relieving throat and bronchial irritation. Camphor is also used as a decongestant and as an essential oil in aromatherapy.

Modern applications of camphor also include its use as an insect and moth repellent to protect natural fabrics from damage, as an anti-microbial substance, hence its effectiveness as a preservative in embalming and as a component in fireworks. In an enclosed space, solid camphor is in equilibrium with its vapor, due to its ability to sublime. In this way, a little camphor kept in tool chests will protect tools against rust as a thin film will sublime on a cold surface.

EXTRACTION OF CAMPHOR BY STEAM DISTILLATION

The presence of the single dipolar carbonyl group in an otherwise saturated cyclic molecule makes camphor only slightly soluble in water, a liquid composed of dipolar molecules. In fact, camphor may be regarded as immiscible with water. It is this physical property, which allows camphor to be extracted by a process known as steam distillation (Figure 6.2).

Camphor oil is isolated both in the laboratory and on an industrial scale by passing steam through the pulverized wood chippings of the camphor tree and condensing the vapors that arise. Camphor crystallizes from the oily part of the distillate and is then purified by sublimation (see Glossary).

Steam distillation is absolutely necessary because any attempt to distill camphor directly from the raw organic material would cause the camphor oil to decompose under the heat needed to bring it to the boil.

Due to the fact that camphor oil and water are immiscible, each substance in the mixture vaporizes independently. It is well known that a liquid boils when the pressure of its vapor equals atmospheric pressure. Dalton's empirical law of partial pressures reminds us that the total pressure of a mixture of gases (or vapors) is equal to the sum of the partial pressures of the individual gases (or vapors). Effectively, the vapor pressures of each liquid in an immiscible mixture add together.

Hence, a mixture of two immiscible liquids will boil at a temperature lower than the normal boiling point or either component in the mixture (100°C and 209°C respectively for water and

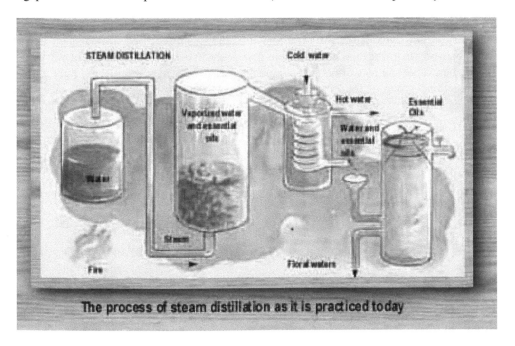

FIGURE 6.2   Illustration of the process of steam distillation.

camphor). Because the vapor pressure of water is much greater than the vapor pressure of camphor oil, the mixture will boil at a temperature only just below the normal boiling point of water causing the camphor oil to vaporize under moderate conditions leaving the substance intact.

THE CARBONYL GROUP AND NUCLEOPHILIC ADDITION REACTIONS

The carbonyl group in a ketone (alkanone) and in an aldehyde (alkanal) is planar. It is also weakly dipolar and, therefore, the double bond can undergo nucleophilic addition reactions.

A very useful step in organic synthesis is the addition of a carbon atom to a molecule.

$$CH_3.CO.CH_3 + HCN = CH_3.COH.CH_3.CN$$

An aqueous solution of hydrogen cyanide can be refluxed gently and carefully with a ketone in a fume cupboard, given the poisonous nature of hydrogen cyanide. The carbonyl bond is attacked by cyanide anions to produce an intermediate substance with a nitrile functional group, which can then be converted readily to other compounds. The nitrile compound produced in solution is racemic with equal concentrations of each of two enantiomers (see the chapter on 'Saving the Pacific Yew Tree' in Part II for more on chirality).

In another example of a nucleophilic addition reaction involving a carbonyl group, an aldehyde can be reduced to a primary alcohol or a ketone can be reduced to a secondary alcohol by refluxing gently with a solution of lithium tetrahydroborate ($LiBH_4$) in a solution of water and methanol.

$$CH_3.CHO + 2H = CH_3.CH_2.OH$$

$$CH_3.CH_2.CO.CH_3 + 2H = CH_3.CH_2.CH_2 OH.CH_3$$

The solution contains a rich source of hydrogen anions, which now attack the slightly positive carbon atom of the dipolar carbonyl bond.

INFRA-RED SPECTROSCOPY AND THE DETERMINATION OF THE STRUCTURE OF ORGANIC MOLECULES

Infra-red spectroscopy exploits the infra-red region of the spectrum of electromagnetic radiation where the frequency is lower than in the visible part of the spectrum and associated wavelengths are longer. When vapor samples of organic compounds are irradiated with infra-red light, different bonds and functional groups in the molecule absorb energy in a narrow range of frequencies specific to them. The absorption of energy causes these bonds to vibrate with greater frequency. The drop in intensity in the infra-red radiation at certain frequencies is detected by a spectrometer and is used as a 'fingerprint' to identify the functional groups or bonds present providing improved understanding of the molecular structure of the compound under examination (Figure 6.3).

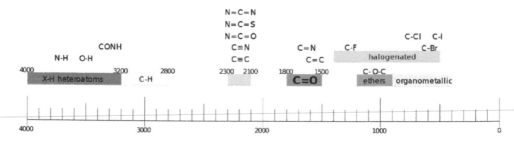

FIGURE 6.3 Infra-red absorption frequencies (in wavenumbers) of various functional groups.

Source: Faverik. Creative Commons Attribution Share Alike International License

In this way, the technique of infra-red absorption spectroscopy assists greatly in the elucidation of the structure of an organic substance.

ATMOSPHERIC ABSORPTION OF INFRA-RED RADIATION; GREENHOUSE EFFECT AND GLOBAL WARMING

The sun emits a wide spectrum of electromagnetic radiation (Figure 6.4), which includes the infrared (wavelength 2,500–700 nm), visible (wavelength 700–400 nm), and ultraviolet (wavelength 400–280 nm). One nanometer (nm) is one thousand billionth of a meter.

Much of the radiation from the Sun reaching the Earth (also called insolation) is absorbed in the higher reaches of the atmosphere by gases, free radicals and ions.

Some radiation is directly reflected back into Space from the Earth's surface by snowfields, by water in the oceans and by clouds while some high energy radiation is absorbed and re-emitted at a lower energy in the infra-red. Collectively, the radiation escaping the Earth into space is known as the albedo.

However, some of the direct and returned radiation is absorbed in the lowest, densest part of the atmosphere by certain atmospheric gases and is re-emitted in all directions including back to the Earth. Infra-red radiation is strongly absorbed by water vapor, carbon dioxide, methane and nitric oxide and nitrous oxide (pollutants from vehicle exhausts), which are present in the atmosphere of the Earth. In turn, some of this absorbed energy (increased vibration in chemical bonds) is passed to other molecules, which are not active in the infra-red through collision. The average translational or kinetic energy of molecules in the atmosphere therefore increases and the temperature rises.

If the amount of these gases increases in the atmosphere disproportionately, as a direct result of human activity, then the heating effect created makes a contribution to global warming— hence, the coining of the term greenhouse gas (Figure 6.5). Average global temperature has risen steadily since the start of the industrial revolution when the extraction and burning of fossil fuels began in earnest.

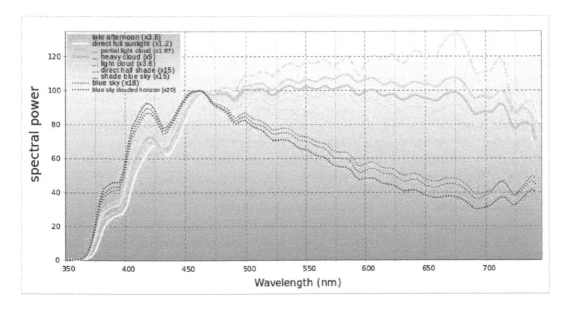

FIGURE 6.4 Spectrum of solar radiation.

Source: available from Wikipedia Commons and licensed for release under the Creative Commons Attribution-Share Alike 4.0 International

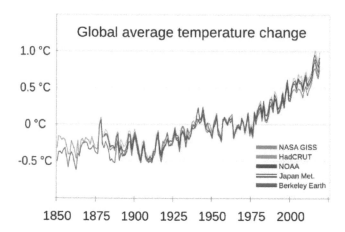

FIGURE 6.5   Global Average Temperature Change compiled from different worldwide sources as of March 24, 2020.

Source: Creative Commons Attribution-Share Alike 4.0 International license

Hydrocarbon bonds absorb particularly strongly in the infra-red in comparison with the double bonds in carbon dioxide and the hydroxyl bonds in water vapor so small, inadvertent emission of gaseous hydrocarbons is very significant. Yet contributions to a rising level of hydrocarbons in the Earth's atmosphere can occur in the most innocent of ways—even in agriculture. Cattle reared for beef or milk production release methane as do the wet paddy fields where rice is grown. As more land is brought into cultivation in a populous world, so methane emissions rise. Ways of limiting global warming are discussed in the chapter on 'Our World of Green Plants—Human Survival' in Part VII.

The mean temperature of the atmosphere, the oceans and the surface of the Earth increases by almost imperceptible levels due to release of gaseous hydrocarbons (especially methane). These gases are products of the decay of dead plants and organisms that are released from the softening, partly frozen peat in the tundra and from sediments in the deep ocean. Methane is not soluble in water and enters the Earth's atmosphere, thereby accelerating the greenhouse effect.

In 2005, the first international treaty aimed at cutting greenhouse gas emissions came into effect. It is known as the 'Kyoto Protocol' after the Japanese city in which discussions first began in 1997. The protocol seeks to control emissions of six greenhouse gases; methane, carbon dioxide, hydrofluorocarbons, perfluorocarbons, nitrous oxide and sulfur hexafluoride.

Release of hydrocarbons into the atmosphere also causes damage to the ozone layer in the upper atmosphere. This was the subject of the 'Montreal Protocol' in 1989 through which international agreement was reached to phase out, by the year 2000, use of hydrofluorocarbons particularly as they are also greenhouse gases. Hydrofluorocarbons had, until that time, been utilized as the propellant gas in aerosol cans and as the coolant material in refrigerators. It is encouraging to note that the Montreal Protocol has been reasonably successful having been largely observed internationally by governments and business who have sought practical alternatives to the continued use of hydrofluorocarbons.

Chemical pollution of the Earth's atmosphere by other means, through the release of the combustion products of organo-sulfur compounds for instance, is presented in the chapter on 'Garlic and Pungent Smells' in Part IX.

ROYAL BOTANIC GARDENS, KEW, UNITED KINGDOM

The Royal Botanic Gardens at Kew in southwest London is justifiably renowned for its extraordinary catalogue of plants and for its success in plant research and conservation. In response to continuing changes in plant habitat throughout the world brought about by climate change, the management of the Royal Botanic Gardens has invested heavily in the creation and development of a new facility, The Millennium Seed Bank, at Wakehurst Place in Sussex, England. This ambitious and visionary program collects seeds from all corners of the globe and stores them for posterity in carefully controlled conditions. In particular, global warming is gradually denying plants which live in delicate or marginal environments a stable and suitable habitat. Examples of such environments are

- the arctic and sub-arctic tundra
- high mountain chains, the Alps, Andes and Himalayas and
- micro-climates in and around isolated valleys or close to geophysical features such as waterfalls.

The aims of the Millennium Seed Bank are to conserve and preserve the seeds and spores of many, many thousands of species of plants, some of which are endangered, so that their characteristics and properties may be systematically studied both for their own sake and also to see whether chemical extracts from them would be of benefit to mankind.

CARBON DIOXIDE AND GLOBAL WARMING

The concentration of carbon dioxide in the Earth's atmosphere is of urgent concern. It continues to rise reaching well over 400 parts per million (ppm) in 2023. CO_2 levels today are greater than at any point in the past 800,000 years (Figure 6.6). Much publicity has been given to the steady rise in the concentration of CO_2 in the Earth's atmosphere due to the unsustainable practice of burning of fossil fuels, notably coal and oil.

Why is so much attention given to the amount of CO_2 in the atmosphere, when a number of other naturally occurring gases, such as methane, are intrinsically more powerful and also contribute to global warming? A remarkably strong correlation exists between global temperature and the concentration of carbon dioxide in the atmosphere (Figure 6.6). The concentration of CO_2 in the atmosphere over time is known from the study of ice cores drilled and extracted from the deep ice sheet in Antarctica. NOAA is the National Oceanographic and Atmospheric Administration funded by the government of the USA. The concentration of CO_2 in the atmosphere has shot up to unprecedented levels since the start of the industrial revolution in the 1850s.

Concentrations of CO_2 (only 0.04% of the atmosphere) and CH_4 (in trace amounts measured in parts per million) do not appear to be significant at first sight. However, CO_2 and CH_4 are the principal greenhouse gases in that order of importance, and both are capable of powerful contributions to the greenhouse effect. However, two other properties of these gases need to be considered in addition to their concentration in the Earth's atmosphere;

- how long they stay in the atmosphere (their lifetime)
- how readily they absorb energy in the infra-red (their radiative efficiency).

CO_2 remains in the climate system for a very long time and so recent emissions produce increases in atmospheric concentration of CO_2 that will persist for thousands of years. This is why the notion of carbon capture (carbon sequestration) by removal of CO_2 from the atmosphere is so important. While

CO$_2$ during ice ages and warm periods for the past 800,000 years

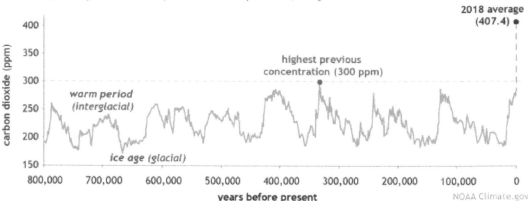

FIGURE 6.6 Based on data from the EPICA Dome C in the Antarctic provided by the NOAA NCEI Paleo-Climatology Program.

CH$_4$ emitted today will only last about a decade, it absorbs much more energy than CO$_2$. In fact, on the basis of a direct comparison of the effect of molecules, CH$_4$ is 25 times more potent than CO$_2$ since it is a tetrahedral molecule rather than a linear molecule with more freedom to vibrate and hence absorb and re-radiate infrared energy. CH$_4$ also has some indirect effects too in that it is involved in reactions in the troposphere (lower atmosphere) that produce another noxious greenhouse gas, ozone.

Questions

1. Examine the molecular structure of camphor presented earlier in the chapter. There are three isomers of camphor, two of which rotate plane-polarized light. Can you identify each one?

2. When acetone is refluxed with an aqueous solution of hydrogen cyanide, the nitrile compound produced in solution is racemic with equal concentrations of each of two enantiomers. What conclusions can be drawn from this fact about the mechanism of the attack of the nitrile anion, a nucleophile, on the carbonyl double bond? How would the presence of enantiomers be demonstrated in practice?

3. Explain the difference between infra-red radiation and microwave radiation. Explain how the former is used to help identify the make-up of organic molecules and why the latter is used to heat up substances including food and also as a carrier wave in telecommunications.

4. Explain what is meant by the term greenhouse gas. Give a comprehensive account of the various ways in which the activities of humankind, directly and indirectly, are introducing more greenhouse gases into the atmosphere of the Earth. Describe potential solutions for reducing the level of emission of greenhouse gases including modified farming practice and resort to alternative sources of power.

5. Explain how infra-red spectroscopy can be used to monitor the concentration in parts per million of greenhouse gases in the Earth's atmosphere.

6. Explain why CO$_2$ is much more influential in climate change than CH$_4$.

7. Chlorofluorocarbons (CFCs) were in widespread use during the 20th century. Explain why the Montreal Protocol banned the industrial and commercial application of CFCs. Give three examples of their use and provide examples of alternative substances that are now in use.

8. Do you consider National Botanical Gardens to be a valuable resource? Explain your reasoning.

SUGGESTED FURTHER READING

A. D. Cross and R. A. Jones. 1969. *An Introduction to Practical Infra-Red Spectroscopy*, 3rd Edition. Springer Science.

K. Nakanishi and P. H. Solomon. 1977. *Infrared Absorption Spectroscopy*, 2nd Edition. Emerson-Adams Press.

BIBLICAL RESINS: FRANKINCENSE AND MYRRH

Abstract: *Frankincense and myrrh, derived from resins of shrubs of Arabia, were reputably gifts to the infant Jesus from the Three Wise Men. Appreciation of essential oils and exotic fragrances spread west from Asia. Trading in the materials of incense comprised a major part of commerce along the Silk Road and other trade routes. Incense refers to any aromatic plant material that releases fragrant smoke when burned. Indeed, the term comes from Latin, incendere, meaning simply 'to burn'.*

Incense is still used for a variety of purposes in the modern world: as an integral part of the spiritual ceremonies of all the main religions, as an air freshener, as an insect repellent and in aromatherapy and meditation. The search for new exotic fragrances and essential oils continues apace today.

Organic Chemistry

- *essential oils*
- *steric hindrance*
- *nomenclature of cyclic terpenes.*

Context

- *cyclic terpenoid acids*
- *cyclic terpenols.*

SOURCES AND USES OF FRANKINCENSE AND MYRRH

The scarcity of frankincense and myrrh meant that at one time they were more valuable than gold. This led to a continuing quest for frankincense and myrrh as exotic fragrances and for incense. Through its value in spiritual life, ancient civilizations retained deep folkloric belief in frankincense regarded as the male counterpart to the femininity of myrrh. Many religions worldwide still use frankincense in their worship of the divine.

A BIBLICAL JOURNEY AND THE THREE GIFTS

The Magi (also referred to variously as the Three Wise Men, the Three Kings, the Three Astrologers or Kings from the East) were a group of distinguished foreigners who were said to have visited Jesus after his birth, bearing gifts of gold, frankincense and myrrh.

The Magi feature regularly in traditional accounts of the nativity celebrations of Christmas and are an important part of the Christian tradition. Three gifts are explicitly identified in the Bible as gold, frankincense and myrrh: gold as a precious and attractive metal, frankincense as a perfume and myrrh as oil for anointing. The three gifts may also have a deeper meaning: gold, a symbol of kingship; frankincense, a perfumed incense representing spirituality and myrrh, an embalming oil, a symbol of death and Man's short time on Earth.

The scarcity of frankincense and myrrh might have suggested at one time they were more valuable than gold. Their scarcity led to a continuing quest for frankincense and myrrh as exotic fragrances and for incense. Historically, high-quality resin has been produced along the northern coast of Somalia, which was supplied to the Roman Catholic Church

Frankincense has a long history of use in India as an Ayurvedic medicine. Frankincense has anti-inflammatory properties providing relief from the pain of rheumatism and arthritis. In manufacturing, Indian frankincense resin oil and extracts are used in soaps, cosmetics, foods and beverages.

Both frankincense, *Boswellia serrata* and myrrh, *Commiphora myrrha* belong to the family, *Burseraceae*. The frankincense tree (Figure 6.7) is believed to originate from North East Africa and Southern Arabia where it is still found.

The 'weeping' of a gum from the frankincense tree yields a fragrant, aromatic resin (Figure 6.8). A volatile oil distilled from this resin is used in aromatherapy to counteract anxiety and relieve tension. The oil may be applied topically or inhaled.

Frankincense has a long history of use in India where it is highly regarded in Ayurvedic medicine. Frankincense has anti-inflammatory and analgesic properties providing relief from the pain of rheumatism and arthritis. In manufacturing, Indian frankincense resin and extracts are used in soaps, cosmetics, foods and beverages.

Olibanum is another name for frankincense that arises from the milky resin exuded from incisions in the bark of several *Boswellia* species, including *Boswellia serrata*, *Boswellia carterii* and *Boswellia frereana*. The exudate from *Boswellia serrata* is most commonly used medicinally. Frankincense is tapped from the trees, allowing the resin to seep out and harden. The hardened drops of resin are called tears because of their shape. Tapping is performed two to three times a year. Historically, frankincense resin of the highest quality has been produced along the northern coast of Somalia which is supplied to the Roman Catholic Church and used as incense.

A thorny shrub about three meters in height (Figure 6.9), myrrh produces pink and yellow flowers and beaked fruit. Its origin is believed to be Somalia, Ethiopia and Kenya.

The resin (Figure 6.10) has astringent, antiseptic and anti-inflammatory properties (Banno et al., 2006). These properties make it helpful in treating mouth and throat infections and the common cold. In summary, the resin may be used to treat mild inflammation of the oral and pharyngeal mucosa.

In commercial manufacturing, myrrh is applied as a fragrance and fixative in cosmetics. Importantly, myrrh continues to be used in embalming and as incense. Also, the essential oils from both myrrh and frankincense are used in aromatherapy.

FIGURE 6.7   Frankincense tree, *Boswellia serrata*.

Source: www.shutterstock.com—stock number ID:51174910

FIGURE 6.8   Frankincense—a fragrant aromatic resin.

Source: www.shutterstock.com—stock number ID:122765479

FIGURE 6.9   Myrrh tree, *Commiphora myrrha*.

Source: Medizinal Pflanzen published by FE Koehler Medizinal Pflanzen 1895. Public domain 70 years after the death of the publisher (occurred in 1914, now 110 years on)

FIGURE 6.10   Resin oozing from the bark of the Myrrh tree.

Source: www.shutterstock.com—stock number ID:50096023

FRANKINCENSE

The principal, non-volatile constituents of frankincense are α- and β-boswellic acid (see Figure 6.11), both of which have anti-inflammatory properties. The boswellic acids are a group of carboxylic acids consisting of a hydrocarbon 'skeleton' of a pentacyclic triterpene and at least one other functional group. Both α-boswellic acid and β-boswellic acid have a molecular formula, $C_{30}H_{48}O_3$; they are isomers and differ only in their triterpene structure.

The boswellic acids are produced by the *Boswellia* family of plants and make up to 30% of the resin. It should be noted that while boswellic acids are major components of the resin they are not removed by steam distillation process (see also the chapters on 'Camphor' and 'Lavender' in Part VI). These non-volatile components are too large to be carried over with the steam vapor during the steam distillation process.

The extracts of Boswellia are important in Ayurvedic medicine. There are some indications that derivatives of boswellic acid may have the properties of an anti-cancer agent (Su et al., 2011).

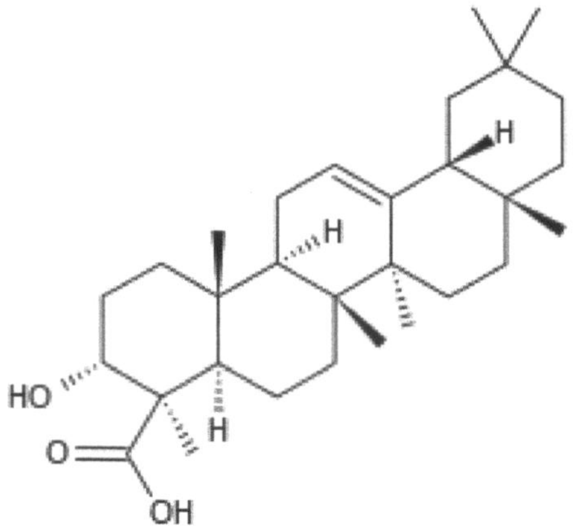

FIGURE 6.11   The structure of α-boswellic acid.

The long-term effects and side effects of taking frankincense have not yet been scientifically investigated. Nonetheless, several preliminary studies have been published and one from researchers at the University of Leicester in the United Kingdom has been included in the references.

By contrast, the lighter terpenes from the resin are much more readily removed with the steam. Consequently, the essential oil collected as the condensate is richer in the lighter molecules and consists of up to 75% of monoterpenes and sesquiterpenes. The most abundant substances are *n*-octylacetate, which is a straight-chain ester derived from octyl alcohol, acetic acid, having the molecular formula $CH_3 (CH_2)_7COOCH_3$ and lesser amounts of 4-ethynyl-4-hydroxy-3, 5, 5-trimethyl-2-cyclohexen-1-one.

MYRRH

The term 'myrrh' derives from Aramaic (murr) and Arabic (mur) meaning 'bitter'. Myrrh has been used throughout history as incense, as a perfume and in embalming and medicine.

The shrub *Commiphora myrrha* exudes an aromatic, yellow resin from its stem known as myrrh. In contrast to the essential oil of frankincense, 4-ethynyl-4-hydroxy-3, 5, 5-trimethyl-2-cyclohexen-1-one (Figure 6.12) is highest in abundance.

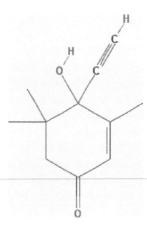

FIGURE 6.12   Major essential oil in the resin of myrrh.

CLASSIFICATION OF CYCLIC TERPENES

Terpenes are polymers and are found mostly in plants (see the chapters on 'Saving the Pacific Yew' in Part II and 'A Plant from the East Indies, Camphor' in Part VI) although larger and more complex terpenes (e.g. squalene) do occur also in animals as steroids.

Terpenes consist of multiples of the isoprene molecule. Isoprene with the molecular formula, $CH_2.C(CH_3).CH.CH_2$, is the name of a branched alkene, which has two carbon–carbon double bonds within a short, five-carbon chain.

Terpenes are classified by use of an empirical feature known as the isoprene rule. Molecules of terpenes, $(C_5H_8)n$, are made up from linear multiples (n) of smaller isoprene molecules, (C_5H_8), which each contain five carbon atoms (where n is an integer).

The naming (nomenclature) is a little confusing:

- where $n = 2$, the molecule is described as a monoterpene
- where $n = 3$, the molecule is described as a sesquiterpene
- where $n = 4$, the molecule is described as a diterpene
- where $n = 5$, the molecule is described as a triterpene and so on

The simplest open chain terpene consists of two isoprene units linked together and has the molecular formula $C_{10}H_{16}$. Terpenes can undergo natural biochemical modification through oxidation or complex rearrangement reactions to produce a variety of open chain and cyclic terpene compounds (see also the chapter on 'Camphor'). In fact, a very wide variety of cyclic terpenes is known. As terpenes, these cyclic molecules are also made up from multiples of the building block, C_5H_8. However, cyclic terpenes do have fewer carbon–carbon double bonds than open chain or acyclic terpenes. A straightforward example of a cyclic monoterpene is the fragrant substance limonene (Figure 6.13), which may be obtained from the rind of lemons.

Returning to the structure of alpha-boswellic acid, shown in Figure 6.11 with the molecular formula $C_{30}H_{48}O_3$, this structure contains five hydrocarbon rings and five terpene units—hence it is described as a penta-cyclic tri-terpene.

STERIC HINDRANCE

One of the most abundant chemicals present in the essential oils of frankincense and especially myrrh is 4-ethynyl-4-hydroxy-3, 5, 5-trimethyl-2-cyclohexen-1-one.

While the extract has an impressive systematic name, it is a cyclic terpenoid having a molecular formula of $C_{11}H_{14}O_2$. Unfortunately, in this case, there is not a more comfortable informal name. However, there is little steric hindrance (see Glossary), which means that the chemistry of the molecule reflects closely the chemistry of the functional groups present.

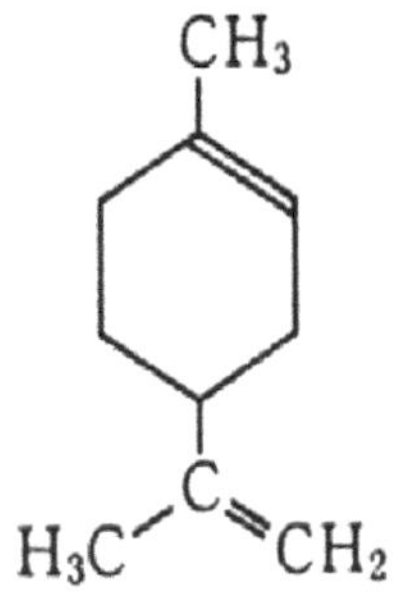

FIGURE 6.13 Structure of limonene.

Steric hindrance can be exploited to change the pattern of the chemistry of a complex molecule by inhibiting an unwanted reaction involving a specific functional group (known as steric protection) or by leading to a preference for one course of stereochemical reaction over another (as in diastereoselectivity).

Questions

1. Why are the boswellic acids much less volatile than other chemical substances present in crude extracts of frankincense? Explain the value of this difference in the steam distillation of the essential oils of frankincense.

2. Give the systematic names for a simple open chain terpene, $C_{10}H_{16}$ and a simple cyclic terpene, limonene.

3. Describe the chemical influences of the alkene, hydroxyl and carboxyl functional groups in boswellic acid.

4. 4-ethynyl-4-hydroxy-3, 5, 5-trimethyl-2-cyclohexen-1-one is a chemical compound present, albeit in different amounts, in the essential oils of both frankincence and myrrh. From the systematic name of the compound, identify the functional groups that are present and give a summary of the principal chemical properties of the molecule.

SUGGESTED FURTHER READING

J. A. Duke. 2008. *Medicinal Plants of the Bible*. CRC Press.

A. Giesecke. 2014. *The Mythology of Plants: Botanical Lore from Ancient Greece and Rome*. Getty Publications.

N. Groom. 1981. *Frankincense & Myrrh: A Study of the Arabian Incense Trade*. Longman.

M. D. Herrera. 2011. *Holy Smoke: The Use of Incense in the Catholic Church*. Tixlini Scriptorium.

G. A. Maloney. 1997. *Gold, Frankincense, and Myrrh: An Introduction to Eastern Christian Spirituality*. Crossroad.

S. Mitchell. 2007. *A History of the Later Roman Empire, AD 284–641: The Transformation of the Ancient World*. Wiley-Blackwell.

B. E. van Wyk and M. Wink. 2004. *Medicinal Plants of the World*. Briza Publications.

REFERENCES

N. Banno, T. Akihisa, K. Yasukawa, et al. 2006. *Anti-Inflammatory Activities of the Triterpene Acids from the Resin of Boswellia Carteri*. J Ethnopharmacol 107: 249–253.

S. Su, T. Wang, J. A. Duan, et al. 2011. *Anti-Inflammatory and Analgesic Activity of Different Extracts of Commiphora Myrrha*. J Ethnopharmacol 134: 251–258.

EUROPEAN LAVENDER

Abstract: *Extracts from the European Lavender find use as a fragrant ingredient in toiletries, cosmetics and detergents; are applied in medicine; and are consumed as a flavoring agent in food and drink.*

Organic Chemistry

- *essential oils, perfumes and cosmetics*
- *separation of essential oils by fractional distillation*
- *hydrolysis and esters*
- *saponification and soap*

Context

- *colloids and hydrosols*
- *vegetable oils and fats*
- *glycerol—a naturally occurring chemical building block.*

EUROPEAN LAVENDER

Lavenders (*Lavandula*) form a genus of 39 species of flowering plants belonging to the mint family, *Lamiaceae*. The genus includes annuals, herbaceous plants and small shrubs. The leaves are long and narrow in most species. The purplish blue color of the common lavender flower is so well known that it has become eponymous.

HISTORICAL NOTE

The ancient Greeks knew lavender as the herb, nard, after the Syrian city of Naarda. Lavender was one of the herbs used in the temple to prepare a holy essence and nard is mentioned in the Song of Solomon. The Greeks discovered early on that lavender would release a relaxing aroma when crushed or burned.

Lavender was used in public baths in Roman times to scent water and it was also considered to be a skin restorative.

The name probably arises from the name of the plant in Latin, lavandarius, from lavanda (things to be washed) and from the verb lavare (to wash).

DISTRIBUTION

Varieties of lavender flourish in dry, well-drained, sandy or gravelly soils and can be found in many parts of Europe, North and East Africa, Arabia and India. The most widely cultivated species of lavender is Lavandula angustifolia. Although native to the western Mediterranean, the plant is commercially cultivated all over the world, including the British Isles, France, the USA, Argentina and Japan (Figure 6.14).

VALUE OF LAVENDER

Food and Drink

Lavender flowers are occasionally blended with black tea, green tea or herbal tea and add a fresh, relaxing scent and flavor. An infusion of flower heads added to a cup of boiling water makes a relaxing beverage.

Flowers can be candied to make 'lavender sugar' and are sometimes used as cake decorations. Lavender is used to flavor baked goods and desserts.

FIGURE 6.14   Fields of lavender in full bloom.

Medicinal Applications

Lavender has had many uses in folk history, a number of which find application today. Infusions of lavender soothe and heal insect bites and burns. Bunches of lavender repel insects. When applied to the temples, lavender oil is believed to moderate headaches. When applied on pillows, lavender seeds and flowers aid sleep and relaxation.

Oil of lavender is an antiseptic and therefore has anti-inflammatory properties. It was used in hospitals during World War I to disinfect floors and walls.

In modern society, lavender is also used extensively in aromatherapy and massage therapy.

Aromatherapy is a form of alternative medicine in which healing effects are ascribed to the aromatic compounds in essential oils.

Cosmetics

The lavender plant, *Lavandula angustifolia*, provides a fragrant oil for balms, salves, perfumes and cosmetics. Lavandin, known as Dutch lavender, also yields an oil, which has a higher level of terpenes such as camphor that add a sharp tone to the fragrance.

Soap and Detergent

The mildly antiseptic nature of lavender and its pleasant, relaxing scent ensure that it continues to be widely used in quality soaps and toiletries.

Essential Oils—Hydrosol from Lavender

An essential oil is an aqueous mixture containing the volatile hydrophobic compounds of an individual plant, which are the 'essence' of its distinctive aroma. Examples of popular essential oils are oil of clove, oil of eucalyptus, oil of peppermint, oil of spearmint, oil of cedar wood and, of course, oil of lavender. The distinctive scent of lavender is due to the presence of key mono-terpenes (Figure 6.15): linalool, a terpene alcohol, and particularly linalyl acetate, a terpene ester.

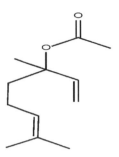 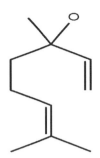

FIGURE 6.15 Linalyl acetate and linalool.

The first step in the preparation of an essential oil from lavender involves subjecting the buds, spikes and flower tips of the plant to steam. The aqueous mixture produced is then fractionally distilled yielding a distillate, which is a mixture of essential oils and water. This mixture of essential oils is known as hydrosol.

Fractional distillation is based on the fact that different substances vaporize at different temperatures at a given atmospheric pressure, thus permitting separation of different parts or fractions of the mixture. Usually, fractional distillation is carried out when the boiling points of the fractions are in a relatively narrow range of temperature, say within 25°C. The vapor will contain higher concentrations of the more volatile components of the mixture and thus the condensate collected will be enriched in the oils. Repeated fractional distillation would be necessary to separate fractions entirely. Typically, these principles are applied on an industrial scale in an oil refinery (Figure 2.6), where the many different fractions in crude oil are separated and also in the cryoscopic fractional distillation of air, which can be liquefied at a very low temperature and high pressure. Cooled liquid nitrogen or oxygen, carbon dioxide and inert gases such as neon and argon are then kept under pressure in cylinders for further use.

Inevitably, the distillate collected from the first fractional distillation of an aqueous liquid mixture will contain various compounds, which vaporize and subsequently condense at or below the temperature set for distillation. These compounds will include water and the distillate will be a hydrosol.

HYDROSOLS AND COLLOIDS

Hydrosols are colloidal suspensions of essential oils in water. A colloidal suspension is formed when microscopic, insoluble particles of a solid or a liquid (the colloid) are dispersed throughout another substance (the continuous phase). The suspended particles are so fine that they do not settle or only do so over a very period of long time. A colloidal suspension is entirely unlike a solution, where the solute dissolves in the solvent to form a single phase. Many colloidal suspensions are translucent because of the scattering of visible light by particles of the colloid. Milk illustrates this effect. It is a colloidal suspension of tiny, immiscible globules of butter fat dispersed in water.

VEGETABLE OILS AND FATS

Vegetable oils and fats are widely distributed throughout the plant kingdom. They are complex mixtures of esters formed from medium to high molecular weight carboxylic acids and the chemical building block, glycerol, which is a trihydric alcohol. The distinction between oils and fats is simply a physical one arising from the difference in the scale of the molecular weight of the hydrocarbon chain; oils are normally liquid while fats are generally solid at ordinary temperatures and pressures.

Glycerol

Glycerol (Figure 6.16) is a colorless, odorless, viscous liquid resembling syrup in consistency. Owing to the polar nature of each of the three hydroxyl groups within a relatively small molecule, glycerol is very hygroscopic and is miscible with water in any proportion. Glycerol is used in the pharmaceutical industry for its softening and moisturizing properties and also as a lubricant. You can see from the figure that glycerol has the functions of both a primary alcohol and a secondary alcohol.

Triglycerides

In vegetable oils and fats, each of the three hydroxyl functional groups is now replaced by a long chain carboxylic acid or fatty acid to make a triglyceride ester. Molecules of natural oils and fats typically involve three different fatty acids. The general formula of a naturally occurring triglyceride is shown in Figure 6.17.

HYDROLYSIS OF ESTERS

Esters may be converted back to their constituent alcohols and carboxylic acids through the hydrolysis reaction. Specifically, a tri-ester of glycerol would be heated in sodium hydroxide solution, a very strong base, producing the sodium salt of the weak acid and glycerol. Addition of a small amount of a dilute solution of an inorganic acid to the reaction products when cool will cause the sodium salt to precipitate as soap. The formation of soap in this way using an aqueous solution of an alkali has led to the hydrolysis of any ester being termed, saponification.

SOAP

The soap produced by hydrolysis of long-chain esters is scented with the addition of an essential oil, which is often oil of lavender containing linalyl acetate. As an example, palm oil was used as the basis of the quality soap products made by the Lever Brothers' company in the United Kingdom (now known as Unilever), which marketed 'Sunlight' soap. In 1864, Caleb Johnson founded a soap company called B. J. Johnson Soap Co. in Milwaukee. In 1898, this company introduced a soap made of palm and olive oils marketed as 'Palmolive'. Later, the American company Colgate-Palmolive acquired the brand.

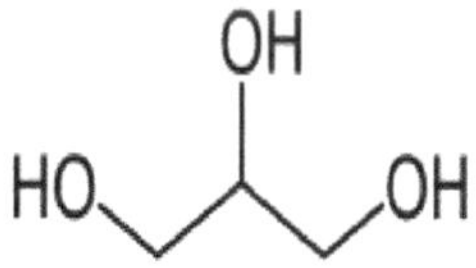

FIGURE 6.16   Glycerol, propane-1, 2, 3-triol.

$$R.COO.CH_2$$

$$|$$

$$R'.COO.CH$$

$$|$$

$$R''.COO.CH_2$$

FIGURE 6.17   The general formula of a triglyceride.

Palm oil is a complex mixture of fatty acids, which include palmitic acid, $CH_3 (CH_2)_{14}COOH$ and stearic acid, $CH_3 (CH_2)_{16}COOH$. Oil palms are grown commercially in huge plantations in the tropics and especially in the East Indies. Palm oil is now used rather more extensively worldwide in processed food, for cooking oil and as a biodiesel fuel. However, the scale of production of palm oil has led to concern about negative environmental impacts, including:

- deforestation
- loss of habitat for endangered species such as the orangutan and Siberian tiger
- potential for climate change through greater 'greenhouse' gas emissions though this is moderated to some degree through the absorption of carbon dioxide by the growing palms.

How Soap Works

Soap is a detergent. The term applies to any material that facilitates removal of dirt from the surface of something to be cleaned. In the modern world, there are many man-made materials other than soap that are detergents.

Carboxylic acids and salts having alkyl chains longer than eight carbons exhibit unusual behavior in water due to the presence of both hydrophilic (carboxylic) and hydrophobic (alkyl) regions in the same molecule. Fatty acids made up of ten or more carbon atoms are nearly insoluble in water, and because of their lower density they float on the surface of water. These fatty acids spread evenly over the water surface forming a very thin film, a monomolecular layer, in which the polar carboxyl groups are hydrogen bonded at the water interface and the hydrocarbon chains are aligned together away from the water. Substances that accumulate at water surfaces and lower the surface tension are called surfactants. By reducing the surface tension of water, surfactants allow water as the cleansing solvent to wet a variety of materials.

The alkali metal salts of fatty acids, as soaps, are more soluble in water than the acids themselves and are strong surfactants. For example, sodium stearate,

$CH_3 (CH_2)_{16}COONa$, is a soap with an anion attached to a long, non-polar, alkyl stem. In small concentrations, soap as a surfactant will dissolve in water producing a random dispersion of solute ions. However, when the concentration is increased, an interesting change occurs. The surfactant ions assemble in spherical aggregates called micelles. The hydrophobic chains of the ions form the center of the micelle and the polar carboxylic ions extend into the surrounding water where they participate in hydrogen bonding.

Another problem relieved by soap relates to the natural presence of dissolved calcium and magnesium salts in the water supply (hard water). The metal cations are removed as an insoluble precipitate of the salts of carboxylic acids, which is commonly recognized as scum.

In summary, the presence of soap in water facilitates the wetting of all parts of the object to be cleaned, removes water-insoluble dirt by incorporation in micelles and softens water by removing dissolved calcium and magnesium salts as scum.

Questions

1. What is meant by the term 'essential oils'? Explain the physical difference between oils and fats.
2. Give an account of the principles lying behind the fractional distillation of a mixture of organic liquids. Show how fractional distillation may be done in the laboratory and is carried out in industrial applications. Also, explain how the process of cracking in an oil refinery is different from fractional distillation.
3. Account for the physical and chemical properties of glycerol that are dominated by the three hydroxyl functional groups present in the molecule.

4. Compare and contrast the properties of vegetable oils (esters of fatty acids), obtained directly from plants, with those of mineral oils (alkanes), obtained from deposits of crude petroleum.
5. Palm oil is used extensively worldwide in processed food, for cooking oil and as a biodiesel fuel. Discuss in a balanced way the arguments in favor of growing of crops for biofuel in relation to environmental concerns such as deforestation and the effect of net 'greenhouse' gas emissions on climate change.
6. Explain how soap acts as a detergent in water.

SUGGESTED FURTHER READING

S. M. Cavitch. 1994. *The Natural Soap Book*. Storey Publishing.
S. Festing. 1982. *The Story of Lavender*. Borough of Sutton, Libraries and Arts Services.
M. Lis-Balchin. 2002. *Lavender; The Genus Lavandula*. Taylor and Francis.
J. Rose. 2007. *Hydrosols & Aromatic Waters*. Institute of Aromatic & Herbal Study.

GLOBAL ALOE

Abstract: *The aloe leaf has been adopted as the symbol of the Royal Society of Veterinary Medicine, UK. Recorded in the Bible, used by the ancient Egyptians, fought over by Alexander the Great, this African plant was traded by the conquering Spanish army in South America and is also found in Traditional Chinese Medicine. Aloe has many important medicinal properties, which improve human and animal health.*

Organic Chemistry

- *cosmetic oils*
- *chromatography—thin layer, gas and gas liquid*
- *ion exchange properties*
- *sustainable practice and the extraction of hydrocarbons by 'fracking'.*

Context

- *polysaccharides and glycosides*
- *aloin and emodin.*

ALOE VERA

Aloe vera (Figure 6.18) is a short-stemmed succulent plant growing fifty centimeters to one meter tall. The leaves are thick and fleshy, green to grey-green, with some varieties showing white flecks on their upper and lower stem surfaces.

The plant has been used in herbal medicine since the beginning of the 1st century AD. Carvings over 6,000 years old have been found in Egypt that contain images of the plant, referred to as the 'plant of immortality'. It was given as a burial gift to deceased pharaohs. Furthermore, early records of the use of *Aloe vera* appear in various ancient texts including the Ebers Papyrus from the 16th century BC, in Dioscorides' *'De Materia Medica'* and Pliny the Elder's *'Natural History'* written in the middle of the 1st century AD.

ALOE VERA AND SKIN TREATMENT

Extracts from *Aloe vera* are used widely in traditional herbal medicine in many countries. For example, *Aloe vera* in Ayurvedic medicine is used as a multipurpose skin treatment. Today,

FIGURE 6.18   Aloe vera.

extracts from *Aloe vera* are widely used in the cosmetics and alternative medicine industries. It has rejuvenating, healing, soothing properties and offers a moisturizing effect especially in cosmetic formulations where it can also provide relief from sunburn (Eshun and He, 2004). The effect is due to the presence of polysaccharides in the extracts (also see the chapter on 'An Asian Staple—Rice' in Part III for more on polysaccharides).

ALOE VERA AND POLYSACCHARIDES

Monosaccharide sugars may be linked through glycoside linkages to form polysaccharides. Glycosides contain a molecule of a sugar, which is bound to another functional group via a glycoside bond. A good example of a glycoside is aloin, found in *Aloe vera*—described later in the chapter.

Glycosides play numerous important roles in living organisms. Many plants store sugar in the form of inactive glycosides, which can be broken down by hydrolysis, promoted by an enzyme when the sugar is needed for use.

An important polysaccharide in the natural world is cellulose. Cellulose has the formula $(C_6H_{10}O_5)_n$ where n ranges from 500 to 5,000, depending on the source of the polymer. Over half of the total organic carbon in the earth's biosphere is in cellulose found for example in cotton and plant fibers. The wood of bushes and trees contains roughly 50% cellulose. See the chapters on 'An Asian Staple—Rice' in Part III for more on the glycoside link and polysaccharides and 'The Wonderful World of Wood' in Part VIII for more on cellulose.

GALACTOMANNANS

The galactomannans form an important group of polysaccharides derived from aloe. These polymers are made up from the monosaccharide sugars, mannose and galactose, hence the name of the class (Figure 6.19). The spine of the polymer is essentially mannose with galactose attached as the side group. As a polysaccharide, galactomammans do not impart a sweet taste and do not readily dissolve in water in contrast to mono and disaccharide sugars. It should be noted in discussing these monosaccharide

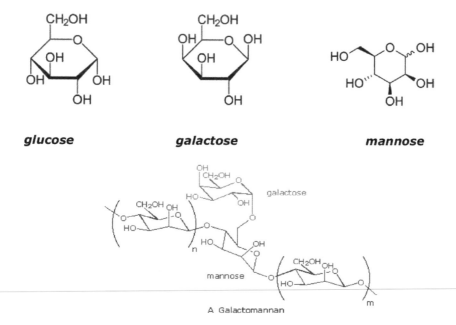

FIGURE 6.19 Three building blocks of sugars: glucose, galactose and mannose and an example of a polysaccharide structure, galactomannan. Note the sugars in blue constitute the linear backbone and the red galactose forms one of the side chains.

sugars that glucose, mannose and galactose are isomers. Each has exactly the same molecular formula and molecular weight but differs only in the configuration of their respective hydroxyl groups in the ring.

Galactomannans are also present in guar gum, which is a powder ground from the guar bean. The guar bean, *Cyamopsis tetragonoloba*, is a legume. Nodules on the roots of the bean fix nitrogen directly from the air due to the presence of bacteria and thereby increase the fertility of soil. Traditionally, the leaves and beans have been utilized not only as animal feed but also as a vegetable for human consumption. Historically, much of the world's supply of guar gum comes from India and Pakistan where the bean is cultivated.

When guar gum is added to water a thick gel is produced. The food processing industry takes full advantage of this property. Galactomannans are added to food products (such as ice cream, salad cream, sauces, tomato ketchup, soups and yogurt) to increase viscosity and hence improve texture.

Guar gum is also used for its medicinal properties as a mild laxative to help relieve chronic bowel conditions, such as colitis and Crohn's disease.

Hydraulic 'Fracking'

A much more recent application of guar gum has sent its market price rocketing. The product has become important as an essential material in drilling for oil and natural gas through the process called hydraulic fracturing, known in common parlance as 'fracking'. In recent times, demand for guar gum has increased to such a degree that farmers formerly living in poverty in north western India and Pakistan have reaped a windfall, which has transformed their lives. Given that the price of guar gum has reached new heights, cultivation of guar beans is now spreading to other semi-arid parts of the world including southern states of the USA such as Texas and Oklahoma. In hydraulic fracturing, a drill hole is initially bored into the earth vertically. At a pre-determined depth, the drill bit and the flexible drill line are then maneuvered into a horizontal position as drilling continues. A mixture of water and chemicals, including guar gum, is pumped down under enough pressure to release the oil and natural gas trapped by fracturing the soft shale rock in which it is contained. The pressure then forces the oil and gas through fissures to the surface where it is collected. The addition of guar gum to the water markedly increases the viscosity of the liquid making high pressure pumping efficient and the process of fracturing the rock more effective (Figure 6.20).

The amount of natural gas made available in this way may be great enough to offset significantly part of the energy demand of a developed industrial nation. Worldwide, it appears that reserves of

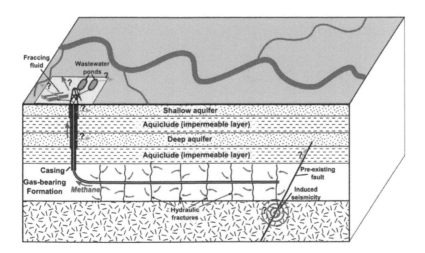

FIGURE 6.20 Schematic depiction of hydraulic fracturing for shale gas.

Source: Mikenorton. Cretive Commons Attribution Share Alike 3.0 Unported license

natural gas held in shale rock could be considerable, which has re-invigorated debate about the ethics of exploitation of further reserves given concern about a greater degree of global warming through the accidental release of methane and other gaseous hydrocarbons and the actual release of their combustion product, carbon dioxide.

ANTHRAQUINONES FOUND IN *ALOE VERA*

The anthraquinones are a common family of naturally occurring substances with yellow, orange and red pigmentation. The chapter on 'Red Dyes; Henna, Dyer's Burgloss and Madder' in Part VII provides much more detail on anthraquinones and on their application as dyes.

Aloin

Aloin (Figure 6.21) is an anthraquinone glycoside. It is extracted from *Aloe vera* as a mixture of two diastereomers, described as aloin A and aloin B. The reference to glycoside means that the polymer has an anthraquinone skeleton (the aglycone), which is then bound to a sugar molecule. In natural products chemistry, compounds possessing a sugar link are known as glycosides. Hydrolysis will generate the aglycone moiety and the free sugar moiety.

Aloin has been used as a traditional medicine since antiquity. It can be used in small amounts as a laxative. The juice of *Aloe vera* has been marketed to support digestive health, although there is neither scientific evidence nor regulatory approval to support this claim. Aloin was a common ingredient sold in over-the-counter laxative products in the United States. However, in 2002, sales were prohibited by FDA due to a lack of convincing safety data.

Aloin is a bitter, yellow-brown colored compound found in species of *Aloe*. Aloin is usually prepared by extraction from aloe latex that seeps out from just underneath the skin of aloe leaves. The latex is then dried and powdered to make the final product.

Emodin

The anthraquinone emodin (Figure 6.22) is another purgative, which can be obtained from the leaves of *Aloe vera* and, incidentally, also from the rhizome and stem of the rhubarb plant.

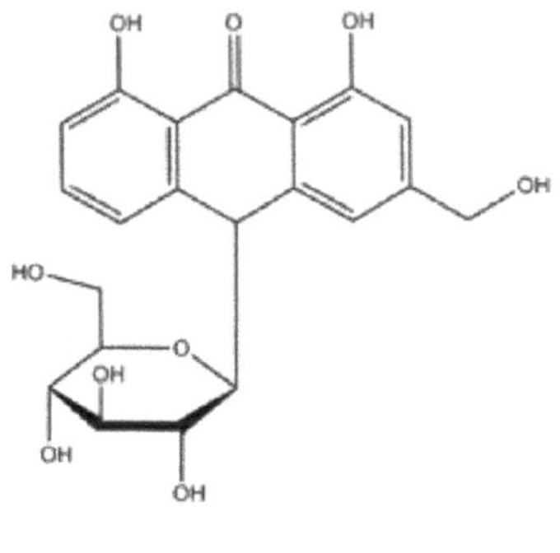

FIGURE 6.21   Structure of aloin.

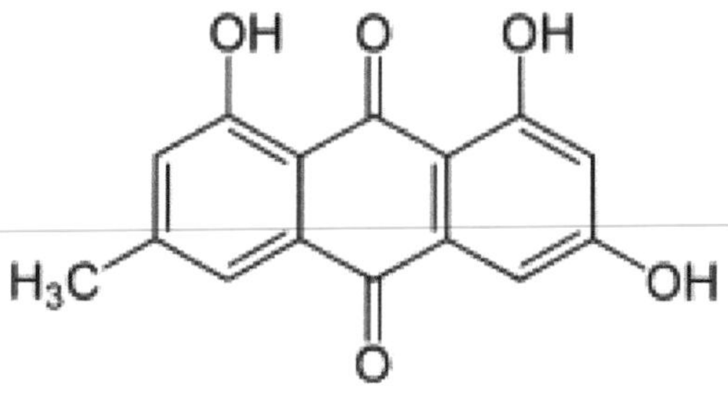

FIGURE 6.22   Structure of emodin.

The chapter entitled 'Red Dyes; Henna, Dyer's Bugloss, Madder' in Part VII provides more on anthraquinones and on the use of emodin as a dyestuff.

CHROMATOGRAPHY

Chromatography is a widely used method for separating and identifying samples of organic compounds, including polysaccharides and long chain isomers, which are structurally similar. An important technical application of chromatography occurs in the pharmaceutical industry where the approach is used for analyzing the purity of drugs. Although chromatography is carried out in different formats (such as thin-layer, column, gas, high pressure liquid, ion exchange), each technique depends upon common basic principles.

There are two essential elements:

- a stationary phase and
- a mobile phase.

The mobile phase is fluid—a liquid or a gas—and contains the various dissolved substances to be separated. The mobile phase moves through or over the stationary phase in contact with its surface. In this way, the dissolved substances in the mobile phase are adsorbed at different rates on the surface of the stationary phase. As a consequence, the substances that spend longer in the mobile phase travel farther over the stationary phase. Separation is achieved.

THIN LAYER AND COLUMN CHROMATOGRAPHY

The stationary phase is a layer of a solid, such as silica gel positioned on a supporting structure of a smooth inert surface such as a plate of glass, hence the name thin layer chromatography. Alternatively the solid material is packed into a vertical column, either loosely or tightly packed under pressure, hence the term column chromatography.

The mobile phase is a solvent, such as ethanol, which passes over the stationary phase by upward capillary action usually in the case of thin layer chromatography and downward under gravity in the case of column chromatography. Different solutes migrate at different speeds and can be separated. The identity of the solutes in the mobile phase is revealed by reference to tabulated records of

- how far they travel over the stationary phase compared to the solvent in thin layer chromatography
- how long it takes (known as the retention time) for the solute to move through the stationary phase in column chromatography.

Thin layer chromatography is often used for separation of small quantities of material under analysis and is a versatile technique for non-polar compounds. The mobile phase, which is also known as the eluate, is commonly a single solvent but could be also a mixture of miscible solvents.

GAS CHROMATOGRAPHY AND GAS-LIQUID CHROMATOGRAPHY

Gas chromatography and gas-liquid chromatography are techniques used for the separation of volatile, non-polar compounds. If the stationary phase is a solid then the process is known as gas chromatography. When oil is used as the stationary phase it is referred to as gas liquid chromatography. A chemically neutral carrier gas, nitrogen or helium, forms the mobile phase, which is passed through a heated, coated column or coil. A film of the stationary phase covers the interior of the long tube forming the coil, which is enclosed by an oven, so that the temperature can be regulated. Substances for separation dissolve at different rates in the stationary phase at a given temperature or remain in the gaseous state in the carrier, the mobile phase. This method has been particularly successful in the separation of volatile substances such as oils, esters of fatty acids and terpenes.

High Pressure Liquid Chromatography

The advantages of high pressure liquid chromatography arise from fast separation and use of small volumes of solvent.

Macro porous resins coated with suitable hydrocarbons are employed as the stationary phase with the size of the pores being selected beforehand to enhance the selectivity of separation. The mobile phase is often a mixture of miscible polar liquids, such as ethanol and water. Applications of the methodology include separation of long chain, structurally related isomers.

Ion Exchange Chromatography

Polar compounds are usually soluble in water. When the compound to be separated possesses an electrical charge as an ion or due to positive or negative functional groups within the molecule, ion exchange chromatography is valuable. The stationary phase is a resin or silicate consisting of large polymeric molecules, often in the form of a cage structure. There are four main types of ion exchange resin differing in the nature of the coating of the functional group, which makes them active:

- strongly acidic (derivatives of sulfonic acid)
- weakly acidic (carboxylic acid groups)
- strongly basic (quaternary ammonium compounds such as trimethyl ammonia)
- weakly basic (primary, secondary or tertiary amines such as polyethylamine).

Two approaches are possible; either the resin is positively charged in which case the mobile phase or elute contains a negatively charged solute or vice versa. In either case, the ionic strength of the mobile phase can be adjusted to alter the retention time.

Questions

1. Give a full account of the application of galactomannans in various industries including food processing, pharmaceuticals, cosmetics, explosives, paper manufacture and textiles.
2. Explain the role of galactomannans in the recovery of oil and natural gas by 'fracking' and give a balanced account of the advantages and disadvantages of the exploitation of naturally occurring reserves of hydrocarbons by this process.
3. Describe fully the principles behind the important practical techniques of different forms of chromatography, which is used so widely in the laboratory and in industry.

SUGGESTED FURTHER READING

R. Cooper and G. Nicola. 2014. *Natural Products Chemistry; Sources, Separations and Structures*. CRC Press, Taylor and Francis Group.
R. H. Davis. 1997. *Aloe Vera: A Scientific Approach*. Vantage Press.
Y. I. Park and S. K. Lee, Eds. 2006. *New Perspectives on Aloe*. Springer.

REFERENCE

K. Eshun and Q. He. 2004. *Aloe Vera: A Valuable Ingredient for the Food, Pharmaceutical and Cosmetic Industries—Review*. Crit Rev Food Sci Nutr 44 (2): 91–96.

Part VII

Colorful Chemistry, a Natural Palette of Plant Dyes and Pigments

INTRODUCTION

Have you ever wondered what makes plants, flowers and fruit so colorful? Why are the leaves green and why do they then turn to wonderful hues of yellow and red in the autumn? The explanation is due to the presence of natural organic compounds that act as pigments.

Chlorophyll is the green pigment of the plant kingdom. The compound gives green plants their characteristic color since chlorophyll absorbs light in the red and blue parts of the visible spectrum of light, while reflecting green and yellow. Importantly, chlorophyll acquires the energy for photosynthesis of carbohydrates from atmospheric carbon dioxide and water vapor with associated release of gaseous oxygen. In the autumn, the leaves lose their chlorophyll, and as the green disappears the remaining natural red and yellow pigments become visible to the naked eye. The bright pigmentation of flowers, fruit and berries attract interest from animals, birds and insects providing food as a reward in return for cross pollination and the dispersal of seed.

Dyes and pigments owe their color to the selective absorption of certain wavelengths of visible light and the reflection of the remainder. Dyes are soluble in water and are usually applied in aqueous solution to a substrate to be colored, such as a textile. In contrast, pigments are insoluble and do not adhere to the surface of the substrate.

Historically, many dye plants were revered for possessing 'magical properties' with the power to heal and to keep evil spirits at bay. Mystery and superstition surrounded the extraction of the essence. An example of this is provided by Woad, *Isatis tinctoria*, which yielded the purple-blue color, indigo, scarce in the natural world.

Plants have a crucial role in sustaining the natural cycles of the planet. Loss of rain forest in the Amazon river system and loss of sea grass on the sandy ocean floor, as just two instances, contribute to diminished air, water and soil quality and to unwelcome climate change, which lead in turn to reduction in the diversity of animal, insect and plant species.

DOI: 10.1201/9781032664927-7

SELECTED READING

M. Pastoureau. 2001. *Blue: The History of a Color.* Princeton University Press.
K. Wells. 2013. *Colour, Health and Wellbeing: The Hidden Qualities and Properties of Natural Dyes.* J Int Colour Assoc 11: 28–36.

A summary of the chemistry in Part VII follows.

Chapter	Organic Chemistry	Context
Green Plants	Photosynthesis	Plant Kingdom
	Chromophores	Chlorophyll
	Chelates and Ligands	Pigments and dyes
	Ultraviolet absorption	Symbiosis
	Spectroscopy	
	Climate change	
	Carbon accounting	
Saffron and yellow dyes	Human health	Carotenoids
	Vitamin A	Carotenes
	Carotenoid dyes	Xanthophylls
	Allyl functional group	Food color and flavoring
Woad and Indigo	Textile dyes	Alkaloid dyes
	Modern synthetic dyes	Indigo, a purple-blue dye
	Azo dyes	Indigo White
Red Dyes, Henna, etc.	Chromophores	Quinone
	Color fastness in dyeing	Naphthoquinones
	Mordants	Anthraquinones
	Applications as dyes	Quinones as building blocks
Reversible Colors	Reversible dyes	Flavonoid dyes
	Color in flowers, leaves,	Carbon accounting
	berries and fruit	Anthocyanins
	Acid/base indicators	
	Circadian and seasonal	Phytochrome,
	rhythms	Photoperiodism

OUR WORLD OF GREEN PLANTS—HUMAN SURVIVAL

Abstract: Directly and indirectly, members of the Animal Kingdom are wholly dependent upon plants. The Plant Kingdom is dominated by green plants converting carbon dioxide and water into a vast and diverse array of organic compounds providing vital resources in food, clothing and shelter and releasing indispensable oxygen into the atmosphere for respiration.

Organic Chemistry

- *photosynthesis*
- *color and chromophores*
- *light absorption and electron delocalization*
- *ultraviolet absorption spectroscopy and the determination of molecular structure*
- *co-ordination compounds (ligands and chelates)*
- *the exploitation of green plants and influence on climate change*
- *carbon accounting and climate change*
- *sustainable practice; natural cycles maintaining air, water and soil quality.*

Context

- *brief notes on the plant kingdom and green plants*
- *lichens and symbiosis*
- *chlorophyll*
- *chlorin, a polymer of pyrrole—a building block in nature.*

HISTORICAL NOTE ON THE PLANT KINGDOM

Robert Whittaker, a distinguished American ecologist, proposed in 1959 a macro-scale classification of life in the natural world into five kingdoms:

- *Animalia; mammals, birds and insects*
- *Plantae; green plants, which include algae and seaweed*
- *Fungi; which obtain nutrition by breaking down organic compounds rather by photosynthesis*
- *Protista; which contain disparate members that are unicellular or multicellular organisms without specialized tissues such as slime molds*
- *Monera; which relate to unicellular organisms such as bacteria.*

For decades, there has been debate about whether fungi may be regarded as plants or animals or something entirely separate from either, but Whittaker's classification appears to have achieved a degree of acceptance in the scientific community (Whittaker, 1978). Members of the Plantae are remarkably diverse and may be broken down into two broad groups:

- *flowering plants and conifers, which reproduce from flowers and seeds*
- *algae, mosses and ferns, which reproduce from spores.*

Lying outside the scope of this book, there are yet more subdivisions of these groups, which derive from consideration of the anatomical structure and function of plants.

GREEN PLANTS

Flowering Plants and Conifers

The importance of green plants to life as we know it cannot possibly be over-stated. Green plants provide most of the world's oxygen, while flowering plants and conifers, numbering well in excess of 250,000 species, furnish much of the food consumed by the members of the animal kingdom, the Animalia, including fruit, berries, nuts, leaves, stems, roots and vegetables. In addition, humankind in particular derives benefit from green plants for clothing, shelter and fuel.

Algae

Algae, singular alga, are a very large and diverse group ranging from unicellular genera, such as *Chlorella* and diatoms forming the phytoplankton of the oceans, to multicellular forms, such as the *giant kelp*, a large brown alga that may grow up to 50 m in length. Most are phototrophic (see glossary) and lack many of the distinct cell and tissue types found in land plants such as stomata, xylem and phloem. The largest and most complex marine algae are called seaweeds, while the most complex freshwater forms are the *Charophyta*, a division of algae that includes *Spirogyra* and the stoneworts. Algae exhibit a wide range of reproductive strategies, from simple asexual cell division to complex forms of sexual reproduction.

Lichens

Lichens are communities of fungi and algae living together in symbiosis where each organism benefits from the presence of the other.

The fungus, a non-green organism, provides structural support to the algal cells covering it and provides nutrients, whereas the algae, green single-celled plants, provide carbohydrates through photosynthesis using its chlorophyll. Lichens can grow almost everywhere in the world from hot desert climates to snow-packed environments but favor moist, temperate zones where they can be found on rocks, roofs, walls and tree trunks. Many lichens have a green or orange tint.

THE VITAL PROCESS OF PHOTOSYNTHESIS

Joseph Priestley, the famous English chemist, showed as early as 1780 that green plants could 'restore air which has been injured by the burning of candles'. He had discovered that green plants produce oxygen.

Later in 1794, Antoine Lavoisier discovered the concept of oxidation but was summarily executed during the troubled times after the French Revolution in 1789 on the grounds that 'he was a monarchist sympathizer and the Republic of France had no need of scientists'!

The overall process of photosynthesis is now well understood and is summarized in the following chemical reaction, which takes place in the presence of sunlight and chlorophyll,

$$6CO_2 + 6H_2O = C_6H_{12}O_6 + 6O_2$$

or more generally as

$$CO_2 + H_2O = (CH_2O) + O_2.$$

Plants then use this glucose either as a direct source of energy or as the building block to drive various metabolic pathways including polymerization to form cellulose for growth or into starch for the storage of energy in the roots or yet again to produce secondary metabolites (see Glossary) for protection.

The vital process of photosynthesis is of paramount importance to animals and insects, which consume plant parts as food and also use them as building materials for shelter.

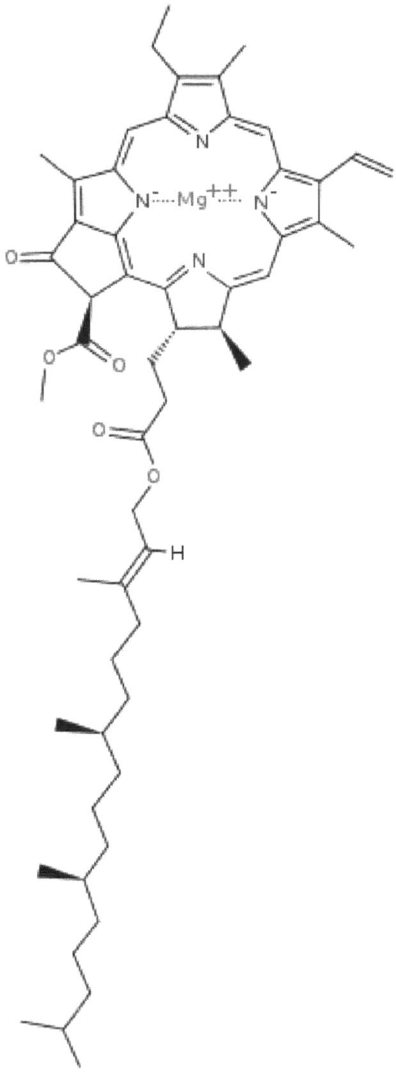

FIGURE 7.1   Molecular structure of chlorophyll A.

CHLOROPHYLL

Chlorophyll represents a family of very large and complex molecules located in the chloroplasts (see Glossary) of the leaves of plants and these compounds are responsible for the characteristic green color. A representative structure of chlorophyll, known as chlorophyll A, is shown in Figure 7.1.

The key naturally occurring building block involved in the structure of chlorophyll is the chlorin ring, which has the ability to co-ordinate with a central metallic ion. In the case of chlorophyll, the central doubly positive cation is that of magnesium. It is known that the magnesium ion plays a very important role in the metabolic pathways of plants.

Chlorin (Figure 7.2) is a particular example of a naturally occurring skeletal unit, known as a porphyrin, which is a polymer made up from essentially four pyrrole molecules (see Figure 5.11 in the chapter on 'Tobacco' in Part V). Chlorin has a large cage-like structure. Incidentally, another example in nature of a complex ring similar to porphyrin is heme, which is a building block in the

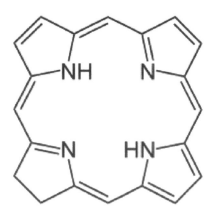

FIGURE 7.2   The cage-like structure of chlorin showing the linkages of four pyrrole groups in each corner.

hemoglobin molecules present in the blood of animals. Four heme groups are bound to a protein called globin. In the case of heme, the metallic ion coordinated is that of iron. In the presence of oxygen, the ionic form of iron is red, whilst in anerobic conditions (see Glossary), namely within the veins rather than the arteries of the body, the ionic form of iron is blue—hence the color of blood. Thus, a co-ordination complex involving iron captures oxygen in the lungs of the animal in a reversible reaction and is instrumental in the chemical process of respiration in animals.

A molecule of chlorin has a cage-like structure with multiple, conjugated, unsaturated bonds: a network of alternating single and double bonds between carbon atoms and a number of nitrogen atoms within the internal space.

Each nitrogen atom has a lone pair of electrons, which makes the internal central space (or interstice) of the chlorin molecule electron-rich. This means that a small positive ion such as a magnesium ion can be incorporated by the chlorin molecule acting as a ligand, the complex ion as a whole being known as a chelate (see Glossary).

Ligands have at least one or more lone pairs of electrons available for bonding. Examples are water and ammonia (one each); ethane-1, 2-diamine (two) and chlorin and heme (multiple). The lone pair of electrons belongs to the oxygen atom or the nitrogen atom in these molecules and is found in the outer valency shell. Since the spin of the two electrons is paired, the lone pair electrons are not normally involved in covalent bonding, unless circumstances arise, as in co-ordination compounds, where the lone pair can be shared with a centrally located ion.

More on the role of co-ordination compounds and ligands in the organic chemistry of dyes can be found in the chapter on 'Red Dyes from Henna, Dyer's Bugloss and Madder' in Part VII. Ligands are also involved in the action of neuro-receptors in the nervous systems of animals and so reference is therefore made to the alkaloids presented in the chapter on 'Maca from the High Andes in South America' in Part IV.

CHLOROPHYLL AND COLOR

Green plants absorb in the blue and red parts of the visible spectrum of light due to two slightly different chlorophyll molecules known simply as chlorophyll a and chlorophyll b. Most of the visible light at 500–600 nm in wavelength is reflected and this is green—hence the color of plants. The reflection of green light is strong enough to mask the effects of weaker absorption of light by carotene (see the chapter on 'Saffron and carotenoids—yellow and orange dyes' in Part VII). When chlorophyll is withdrawn by the plant in the autumn, the leaves have yellow, brown and red tints because the carotene absorbs more weakly in the blue part of the spectrum while reflecting the 'warmer' colored light (Figure 7.3).

FIGURE 7.3 Natural pigments displayed in an autumnal scene at the National Arboretum, Gloucestershire, United Kingdom.

Source: courtesy JJ Deakin

When green plants are cooked for food, ions of magnesium (Mg2+) in molecules of chlorophyll are replaced by those of hydrogen. Cooking causes the breakdown in coloration (bleaching) because delocalization of electrons is facilitated by the central, chelated metal ion.

CHROMOPHORES

A chromophore is the part of a molecule responsible for its color. The color arises when a molecule absorbs certain wavelengths of visible light and transmits or reflects others. The chromophore is a region in the molecule where the energy difference between two different molecular orbitals falls within the range of the visible spectrum. Visible light reaching the chromophore can thus be absorbed in a narrow band of frequencies by exciting an electron from a molecular orbital in the ground state to one in an excited state.

It is important to bear in mind that these electronic transitions are from one molecular orbital to another—these are not atomic spectra.

Chromophores are strongly associated with conjugated unsaturated systems of covalent bonds within molecules. In Part VII on color in organic chemistry, many examples are presented of linear and cyclic molecules: beta-carotene, a linear molecule, in '*Saffron and carotenoids—yellow and orange dyes*'; quinones, cyclic molecules in the '*Red dyes from Henna, Dyer's Bugloss and Madder*' and the purplish-blue dye, indigo in '*Woad and Indigo*'.

The phenomenon of color in chromophores can be understood in terms of the electronic states of molecules. The atoms of the elements carbon, nitrogen and oxygen with atomic numbers 6, 7 and 8 respectively have partially filled p orbitals. The p orbital occupies a space around an atom, which has a 'dumbbell' shape. When a double (or a triple) covalent bond is formed between two of these atoms, the 'dumbell' shapes overlap each other to form a molecular orbital, named a pi orbital. This orbital has a shape resembling a 'sausage' above and below the plane of the bond. In a conjugated system of double and single bonds, the molecular pi orbitals of neighboring bonds merge into one, thus allowing the valency electrons in the p orbitals to range over the system, i.e., the electrons are delocalized (see Figure 7.4).

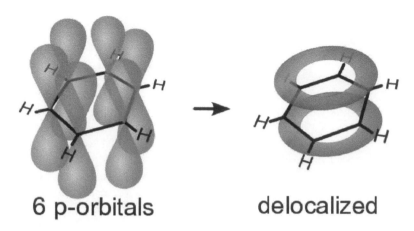

6 p-orbitals   **delocalized**

FIGURE 7.4 Orbital hybridization in benzene rings showing the pi bond.

Source: courtesy of Vladsinger. Available from Wikepedia Commons. Permission granted to copy, distribute and modify the document under the terms of the GNU Free Document License.

Experimental measurements confirm that the carbon-to-carbon bond length in conjugated systems within molecules lies between the length of a single and a double bond. It should be noted that in other parts of this book many different organic substances are presented, with delocalized systems of electrons in molecular orbitals, and each of these substances possesses a great deal of stability.

It has also been found empirically that the longer the carbon skeleton of the chromophore the lower the energy gap between the energy levels of pi orbitals. Thus, the longer chromophores, such as β-carotene, will absorb at smaller frequencies than shorter chromophores like benzene. Benzene and other aromatic substances, such as naphthalene, appear colorless because they absorb in the ultraviolet region of the spectrum at higher frequencies and hence at higher energy than in the visible range. The chromophore of beta-carotene absorbs in the blue end of the visible spectrum, transmitting light in the red and orange areas, which are at lower frequencies, while the chromophore in benzene—and other aromatic substances such as naphthalene and pyridine—absorb outside the visible spectrum in the ultraviolet and, thus, are colorless.

This effect can be seen also when a chromophore is changed to make it smaller. Here it will be instructive to compare the purplish-blue color of indigo with the white color of indigo white in the chapter on 'Woad (Isatis tinctoria) and Indigo' later in Part VII. The color of the new chromophore will be due to light absorption in the ultraviolet at higher frequencies outside the visible region shifting the apparent color perceived by reflected light from blue to white.

Of course, extending the length of a chromophore would have the opposite effect so the inclusion of, for example, an azo functional group between two colorless benzene molecules will form an azo dye (and for more on azo compounds, see the chapters on 'Woad (Isatis tinctoria) and Indigo' and 'Reversible Colors in Flowers, Berries and Fruit' in Part VII).

MOLECULAR INTERACTION WITH ELECTROMAGNETIC RADIATION

Electromagnetic radiation interacts with the molecules of various substances in different ways depending upon the energy of the radiation, the molecular structure of the compound and, to some extent, the physical state of the compound, namely, whether it is a solid, gas or liquid or in solution. The general effects of the absorption of radiation are shown in Table 7.1 where the energy available from radiation increases from top to bottom of the table.

When ultraviolet or visible radiation strikes a molecule of an organic compound the valence electrons in chemical bonds can absorb enough energy to jump to the next energy level. These are

TABLE 7.1

Frequency Ranges of Various Light Sources

Frequency Range	General Effect
Microwave	Translation and rotation of a molecule
Infra-red	Stretching vibrations in bonds
Visible	Color, due to visible light absorption by a chromophore
Near Ultraviolet	Absorption but no color—sometimes associated with subsequent fluorescence in the visible spectrum
Far Ultraviolet	Structural change, dissociation of bonds

quantum steps so the absorption of radiation occurs at a definite frequency relating to the energy gap of the transition involved. Of course, high energy irradiation can cause so much energy to be absorbed that the bonds break to release free atoms or radicals. This occurs, for example, in the case of molecules of ozone in the upper atmosphere of the Earth (see also free radicals in the chapter on 'Cocoa—Food of the Gods' in Part IV).

ULTRAVIOLET ABSORPTION SPECTROSCOPY AND ORGANIC CHEMISTRY

Molecular spectroscopy is a field of study that exploits each kind of electronic transition whether the radiation involved lies in the infra-red, visible or ultraviolet regions of the spectrum. In the spectrum shown in Figure 7.5, one nanometer (nm) is one thousand billionth of a meter. The ultraviolet region lies in the range of wavelength from 400 to 280 nm.

Molecular spectroscopy is of interest to both organic chemists and biochemists who are concerned with the identification of large organic molecules, the identification of key chemical groupings in unknown or little understood organic molecules and with quantitative analysis. Since speed of analysis and the convenient use of small samples are important considerations for these scientists, low resolution ultraviolet absorption spectra in the liquid phase or in solution are especially valuable.

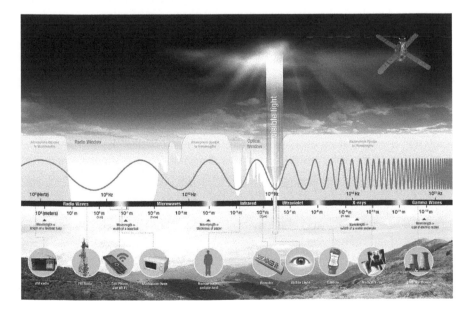

FIGURE 7.5 The electromagnetic spectrum of radiation.

Source: NASA. Public domain

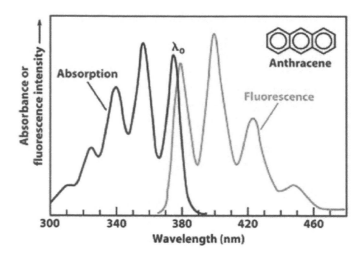

FIGURE 7.6 Absorption and fluorescence of anthracene.

Source: www.reseapro.com Jablonski diagram Archive

The effect of change of solvent has been widely studied but is not that significant, unless hydrogen bonding is involved. The nature of the solvent may slightly vary the absorption frequency of the solute and also the precise shape of the bands.

An illustration of the characteristic shape of an absorption spectrum of ultraviolet radiation is shown in Figure 7.6 for the compound, anthracene. Absorption is followed quickly by loss of the energy through fluorescence within the visible region in a mirror image, which relates to the similar physical structure of the molecule when it is in the higher energy state.

Low resolution ultraviolet spectra have been of great benefit in organic chemistry and complement information gained from studies of the structures of organic molecules arising from infra-red absorption spectra and nuclear magnetic resonance (nmr) absorption spectra.

ULTRAVIOLET ABSORPTION SPECTROSCOPY AND MOLECULAR STRUCTURE

While saturated organic molecules seldom have transitions in the visible or near ultraviolet regions of the spectrum, unsaturated molecules have low-lying excited states that can be reached by absorption of radiation in this region.

The existence of a low resolution visible or ultra-iolet absorption spectrum can even be taken as evidence of unsaturation in molecules. For many unsaturated molecules, the absorption regions are defined and occur at recognizable, typical frequencies for a number of common structures in large organic molecules. There is considerable variation both in the intensity and position of absorption, which is very helpful in identifying them in a given organic compound.

An increase in conjugation—that is alternate single and double bonds—is known to lower the frequency (and also raise the intensity) of the main absorption bands, as seen in Table 7.2, showing a range of unsaturated molecules.

Much is also known about the effect of functional groups on these absorption bands. Functional groups, which do not extend the degree of unsaturation of a molecule, have little effect, though some, including hydroxyl and methyl groups, tend to lower the frequency of absorption in a reliable way. As an illustration, compare phenol with para nitrophenol in Table 7.2

A very large bank of absorption data has been built up over time to help with the identification of an organic molecule or its structural features. Organic chemists regularly measure the ultraviolet absorption spectra of unknown or uncertain compounds (especially in the fields of polyene, aromatic,

TABLE 7.2
Absorption Bands of Some Common Conjugated Compounds

Organic Substance	Wavelength in nm of Peak Absorption	Region
Ethane	180	ultra violet
1.3 butadiene	217	ultra violet
Benzene	255	ultra violet
Phenol	280	ultra violet
Naphthalene	286	ultra violet
Para nitrophenol	320	ultra violet
Anthracene	375	ultra violet
Beta carotene	450	visible

heterocyclic, steroid and carotenoid chemistry) in order to provide evidence for molecular structure. Even where the evidence is not conclusive, it may guide research in the chemistry of these compounds.

GREEN PLANTS, SUSTAINABILITY AND LIMITING CLIMATE CHANGE

Some gaseous substances can absorb energy from the infra-red part of the spectrum, which makes their covalent bonds stretch and vibrate even more. Only molecules comprised of different atoms, which are bonded together and differ in electro-negativity, can interact with infra-red radiation in this way, because of the dipolar nature of the chemical bonds.

Hence, oxygen and nitrogen, the principal natural constituents of the atmosphere by volume (21% and 78% respectively), do not absorb infra-red radiation while carbon dioxide, water vapor, methane and nitric oxide do quite strongly. Gases that absorb in the infra-red are called 'greenhouse gases' because they reduce the albedo effect of the Earth (energy radiated from the Earth back into space).

It is, of course, the balance between incoming radiant energy received from the Sun and outgoing energy radiated from the Earth that fundamentally drives the climate of the Earth so a reduction in albedo results in global warming and climate change. A section of the chapter on 'A Plant from the East Indies, Camphor' in Part VI explores greenhouse gases in more detail.

Simply because of the sheer volume of carbon dioxide released into the atmosphere by man due to the burning of fossil fuels, greater initial attention has been given to measures to control it. However, the inadvertent release of more methane is a threat of some significance because of the very high absorbance of infra-red radiation by hydrocarbons. Some of this additional methane arises from natural sources through the effect of global warming itself on the permafrost in the tundra and on ocean sediments. Yet, more is released by energy-intensive farming practice, which provides protein from animal sources rather than from green plants. A switch in farming methodology toward green plants and a reduction in demand for animal protein arising from a change in diet in wealthy nations of the world would have beneficial effects on climate change.

Welcome and positive contributions to managing the effects of climate change would also arise from an increase in the capture or sequestration of carbon from the atmosphere through increasing photosynthesis by green plants brought about by

- reversal of the scale of de-forestation programs
- reduction in over-grazing of green plant habitat by farmed animals.

Other approaches to carbon capture involve direct physical intervention by man to remove waste carbon dioxide from the exploitation of fossil fuels in the energy generation industry by liquefying the gas under pressure and forcing it into underground reservoirs in deep mines or redundant oil and gas fields.

Reduction in dependence on fossil fuels by the use of renewable sources of energy would be advantageous in respect of climate change though there are distinct practical and political issues associated with this. However, the use of renewable raw materials and the recycling of bio-degradable substances would be of considerable value (UK Government, 2023).

A NOTE ON CARBON ACCOUNTING AND CARBON NEUTRALITY

With comprehensive, binding international agreements on climate change still somewhat elusive, sporadic regional and local initiatives have arisen in enlightened government and business circles.

Though ambitious, some institutions have made public statements about their aim to achieve a zero-carbon footprint or carbon neutrality in the pursuit of their activities. At the very least, such aims offer a strong statement of environmental commitment appealing to many individuals whether electors or customers or those involved in the supply chain.

Carbon neutrality is a condition in which net greenhouse gas emissions associated with an organization or a product are zero. Ideally, in order to make this effective and meaningful, an agreed system of carbon credits would have to be introduced worldwide, which would be typically used to counterbalance or compensate for carbon dioxide emissions elsewhere in a process. Such offsetting would have to be based upon an internationally acceptable unit of measurement or CO_2 equivalent. Very probably, such carbon credits could be legitimately traded between organizations in a regulated market.

However, meaningful, widely accepted standards with which to compare the effects of one human activity with another do not yet exist. There is still no universally accepted unit of carbon accounting although an expression of this is in use. A CO_2 equivalent for a greenhouse gas may be regarded as the amount of CO_2 in metric tons, which would have the same global warming potential when measured over a 100 year timescale.

Although governments in Japan or Sweden for instance have set energy efficiency targets, one clear example of resolve in carbon accounting is provided by the British Government which in 2008 passed the UK Climate Change Act. This act of parliament set up the Department of Energy and Climate Change and established a legal obligation to report annually progress in the UK toward the meeting of carbon targets. In fact, the overall carbon target is to reduce greenhouse gas (ghg) emissions by at least 80% by 2050. Intermediate carbon targets have been set covering the period 2008 to 2027. The UK has made a commitment to halve ghg emissions by 2027 relative to 1990. Where emissions rise in one sector of the economy, the UK government will look to reduce them in other sectors by a corresponding amount. In 2014, emissions were reported to be down by a quarter since 1990. On the basis of current policy, the UK government appears to be on track to achieve a reduction of one third by 2020. However, over the next ten years, technologies will have to be developed and deployed to achieve the overall target of a reduction of 80% by 2050. The UK is attempting to move to a lower carbon, more sustainable economy that is far less dependent on irreplaceable fossil fuels. Much improved fuel efficiency in the space-heating of buildings is expected to make a large contribution.

Currently, almost one third of the population of the world lives in countries where an Emissions Trading Scheme (ETS) is in force. Emissions Trading Schemes (ETS) run by governments remain a key tool in addressing the need to decarbonize to reach the goal of net-zero by the year 2050.

Questions

1. What are lone pairs of electrons and in what atoms are they present in organic molecules? Explain the role of lone pair electrons in chemical bonding. Give examples of linear and cyclic organic molecules that can act as ligands.
2. Provide three examples of symbiosis in the plant kingdom and explain the basis of the relationship between each pair of organisms.
3. Explain why green plants are of critical significance in arresting climate change.
4. Why do green plants appear green?
5. Explain the difference between primary and secondary metabolites in plants and provide examples of each kind of compound.
6. By reference to the theory of molecular orbitals, explain how conjugated systems in organic molecules can give rise to chromophores.
7. Beta-carotene, presented in the chapter on 'Saffron and carotenoids—yellow and orange dyes', quinones, found in the chapter on 'Red dyes from Henna, Dyer's Bugloss and Madder' and the alkaloids, indigo and indigo white, are described in the chapter on 'Woad (Isatis tinctoria) and Indigo'. Explain how the length of the chromophore influences the frequency of light absorbed by the molecular orbital and hence the color of the compound.
8. Account for dipolar bonds in covalent molecules. Many of the gases in the atmosphere of the Earth are transparent to infra-red radiation but two in particular absorb strongly, reducing the albedo effect, and they are known as 'greenhouse gases'. Which are they?
   Explain why many pollutant gases (methane, oxides of nitrogen and carbon hydrogen fluorides) are very influential as 'greenhouse gases' and indicate the main sources of emissions.
9. Give an account of the environmental disadvantages of excessive exploitation of green plants in some parts of the world, examples being deforestation and over-grazing of grasslands. Allow for the fact that, in certain countries, man-made monocultures of crops are grown extensively.
10. What is Zero Carbon Accounting? Give a balanced account of the pros and cons of this worthy and ambitious approach to limiting climate change.
11. What are the environmental effects of the rising level of emissions of carbon dioxide that are being released into the atmosphere of the Earth by the activities of man? Describe natural and artificial means of carbon capture that may address the issue.

SUGGESTED FURTHER READING

C. J. Balhausen. 1962. *Introduction to Ligand Field Theory*. McGraw-Hill.

Emission Trading Schemes. https://icapcarbonaction.com/en/publications/emissions-trading-worldwide-2022-icap-status-report

J. N. Murrell, S. F. A. Kettle, and J. M. Tedder. 1965. *Valence Theory*. John Wiley.

A. Streitwieser Jr. 1961. *Molecular Orbital Theory for Organic Chemists*. John Wiley.

'Zero Carbon Britain: Rethinking the Future'. 2014. Centre for Alternative Technology Eco Store and also at www.zerocarbonbritain.org

REFERENCES

UK Government. 2023. *Reducing the UK's Greenhouse Gas Emissions by 80% by 2050*. UK Department of Energy and Climate Change.

R. H. Whittaker, Ed. 1978. *Classification of Plant Communities, Handbook of Vegetation Science*. Kluwer Academic Publishers.

SAFFRON AND CAROTENOIDS—YELLOW AND ORANGE DYES

Abstract: The golden-colored essence of saffron, namely crocin, has always been a highly prized luxury item fostering trade over millennia and remains one of the most expensive plant commodities by weight.

Organic Chemistry

- *carotenoids which contribute to photosynthesis in a different part of visible spectrum to chlorophyll and to autumn color in plants and are present in tomatoes and root crops such as carrots*
- *human health and vitamin A*
- *retinol and retinal*
- *allyl functional group.*

Context

- *crocin—a terpenoid dye*
- *use in food flavoring and as a food colorant*
- *beta-carotenoid and other terpenoid colorants derived from plants.*

SAFFRON—THE PLANT

The saffron crocus, *Crocus satirus* (Figure 7.7), is unknown in the wild having been developed by selective plant breeding and propagation through vegetative multiplication of the *Crocus cartwrightianus*, which is native to the Greek island of Crete. Documented history of cultivation spans millennia (Rubio-Morago, et al., 2009).

Dried stigmas of the plant are used in food flavoring as seasoning and as a food colorant.

The stigma is the female part of the reproductive system in a flower. It receives pollen and it is on the stigma that the pollen grain germinates. The red structure in the picture is the stigma of the crocus flower.

A NOTE ON SAFFRON EXTRACT—AN EXCLUSIVE LUXURY ITEM TRADED FOR MILLENNIA

Given the labor-intensive methods involved in cultivation, separation of the stigmas and extraction of the golden-colored essence, it comes as no surprise to learn that saffron has always been a highly prized, expensive luxury item, which has fostered trade over millennia. In fact, it may take up to 250,000 stigmas to produce one pound of saffron, therefore saffron remains one of the most expensive plant commodities by weight.

Due to high costs, saffron has been cultivated in the West nearer to its point of use. The cultivation of saffron was introduced into England in 1530 mainly in the drier eastern part of the country. The town of Saffron Walden in the county of Essex owes its name to the plant. Saffron continues to be grown commercially in the eastern county of Norfolk. By 1730, the Pennsylvania Dutch, settlers from central Europe, had established cultivation of saffron in eastern Pennsylvania in the USA, and production continues today around the city of Lancaster. However, today much of the world crop is produced in Iran.

CROCIN

Crocin (Figure 7.7) is a natural xanthophyll carotenoid found in the flowers of the crocus family and is the substance responsible for the golden color of saffron. Crocin is a carotenoid—a colored pigment ultimately derived from terpene. Crocin has a deep red hue but, when dissolved in water, it forms a solution with the characteristic golden orange color. It has a very large molecular mass ($C_{44}H_{64}O_{24}$) and is a crystalline solid at room temperature. The central part of the molecule consists of four isoprene units.

Crocin has been shown to be a potent antioxidant and has also been shown to prevent the proliferation of cancer cells.

FIGURE 7.7   Attractive flowers of the Saffron crocus.

Source: Dobies of Devon, UK, flower and seed merchants

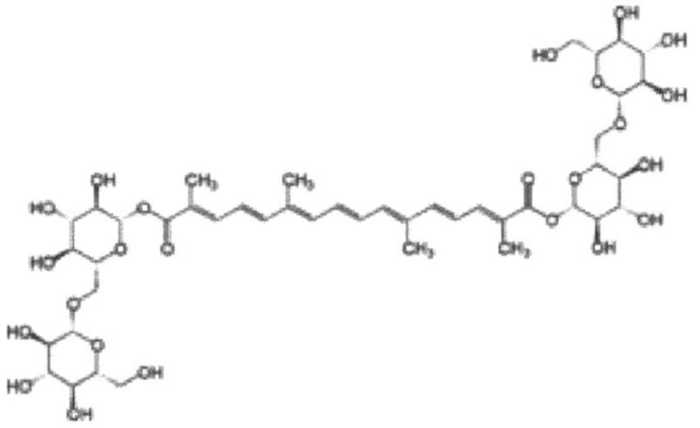

FIGURE 7.8   Structure of crocin.

CAROTENOIDS AND AUTUMNAL COLOR

Carotenoids are pigments ultimately derived from terpene. The breakdown of chlorophyll and other compounds in the leaves of deciduous plants and trees in the autumn, as sugars are recovered and stored over winter in the root system, creates a natural palette of reds, oranges, yellows and browns as carotenoid substances are left behind and exposed.

Without carotenoids in the leaves of plants, episodes of excess light energy could destroy proteins and membranes. In addition, some scientists believe that carotenoids can also act as regulators of developmental responses in plants.

Naturally occurring carotenoids obtained from plants that are used for dyeing include

- lutein from nettles, French marigolds and many other plants
- bixin from annatto obtained from the seeds of the achiote tree (*Bixa orellana*)
- crocin from the stigma of the flowers of the Saffron plant, *Crocus satirus*.

Carotenoids belong to a family of brightly colored natural compounds found in flowers and fruits. They are classified into two families: carotenes and xanthophylls. The former contain carbon and hydrogen atoms, whereas the latter also contain oxygen atom(s) in the structures. All carotenoids have a long hydrogen carbon chain with alternating single and double bonds. The molecular skeleton of carotenoids contains forty carbon atoms arising from eight multiples of the naturally occurring building block, isoprene (see Figure 7.9). They are, therefore, tetra terpenoids.

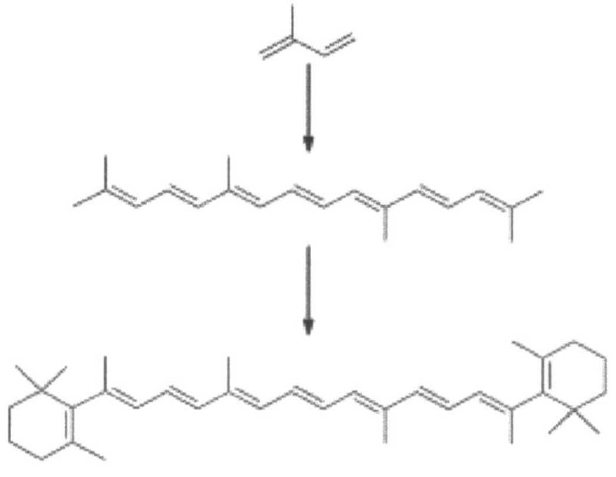

FIGURE 7.9   Beginning from an isoprene building block, the pathway to beta-carotene.

Their color in the visible spectrum of light is due to the large number of double bonds and associated electron delocalization.

CAROTENE CAROTENOIDS

CAROTENOIDS—HISTORICAL NOTE

The earliest studies on carotenoids date from the middle of the 19th century. Carotene was isolated in 1831. In 1837, Berzelius gave the name xanthophylls to the yellow pigments he obtained from autumn leaves. By the 1860s, lutein had been identified. The first separation of the two main types of carotenoids, carotenes and xanthophylls, was achieved through chromatography in 1910 by Willstater and Mieg who established the empirical formula of $C_{40}H_{56}$ for beta-carotene and $C_{40}H_{56}O_2$ for xanthophylls. It was not until the 1930s that Karrer elucidated the chemical structures of carotenoids and also demonstrated that carotenoids are transformed into vitamin A in the human body.

Carotenes are found in nature and carotene-rich foods are often orange in color, e.g., carrots, sweet potatoes, winter squash, spinach, kale and fruits like cantaloupes and apricots.

β-Carotene

β-Carotene (Figure 7.10) is a very important member of the carotenoid family and is the precursor to vitamin A.

Lycopene

Lycopene (Figure 7.11) with molecular formula $[C_{40}H_{56}]$ is a bright red carotene pigment from tomato species and other red fruits and vegetables such as red carrots, watermelons and papayas. Due to eleven conjugated double-bonds, lycopene is deep red and as an unsaturated compound has the properties of an antioxidizing agent. Normally, lycopene is used as food coloring in the food industry.

Isolation and Extraction

Use of supercritical CO_2 is a well-established method for extraction of β-carotene and lycopene from waste tomato paste and even watermelon (for more about supercritical CO_2 see the chapter

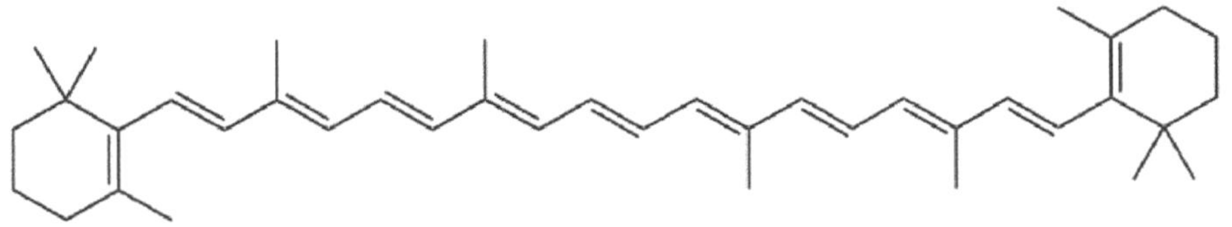

FIGURE 7.10   The chemical structure of beta-carotene.

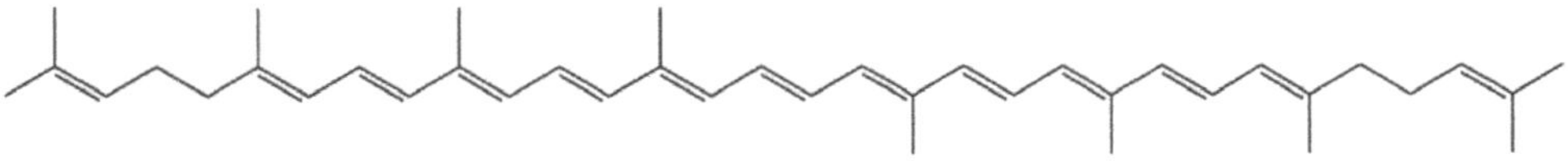

FIGURE 7.11   The molecular structure of lycopene.

on 'Coffee; Wake up and Smell the Aroma' in Part IV). CO_2 is often chosen for supercritical fluid extraction method as it has a low critical temperature, is non-flammable and can be obtained at low cost and in high purity. For example, skins and seeds of dried ground up, ripe tomatoes are extracted as follows: the extraction conditions are 2,500–4,000 psi pressure and 40–80°C for 30 min. Lycopene is more readily dissolved in supercritical CO_2 than carotenes. At 4,000 psi and 80°C, the extraction product contains up to 65% of lycopene and 35% of β-carotene.

XANTHOPHYLL CAROTENOIDS

Lutein

Lutein belongs to the xanthophyll family of carotenoids. Again, the molecular structure is made up from isoprene building blocks in the form of a tetra-terpenoid compound but with oxygen atoms also present (see Figure 7.12). Lutein is obtained from green leafy vegetables such as spinach and kale. Lutein is also responsible for the yellow color of egg yolk, chicken skin and fat. The very name, lutein, comes from Latin meaning 'yellow'.

According to the American Optometric Association, studies show that lutein reduces the risk of chronic eye diseases, including age-related macular degeneration and cataracts.

Age-related macular degeneration (AMD) is a painless eye condition that generally leads to the gradual loss of central vision. The macula is the part of the retina at the back of the eye that is responsible for central vision. As the name of the condition suggests, loss of vision is gradual over many years and usually occurs with ageing. AMD is a leading cause of visual impairment and is most common in people over fifty years of age. It is estimated that one in every ten people over sixty-five has some degree of AMD.

Because of beneficial effects on eye health, small amounts of lutein added to the daily diet are recommended as human beings cannot produce it in the body.

Lutein has one β-ring and one ε-ring. Again, it is the presence of the long conjugated double bonds (polyene chain) which cause its distinctive light-absorbing properties. The polyene chain is susceptible to oxidative degradation by light or heat and is chemically unstable in acids. It is noteworthy that the position of the double bond in lutein involves a chemically reactive, allylic hydroxyl, functional group.

THE ALLYL FUNCTIONAL GROUP

The allyl functional group (Figure 7.13) has the formula $H_2C=CH-CH_2R$ where R represents the remainder of a molecule. The allyl functional group consists of a methylene bridge ($-CH_2-$) attached to a vinyl group ($-CH=CH_2$).

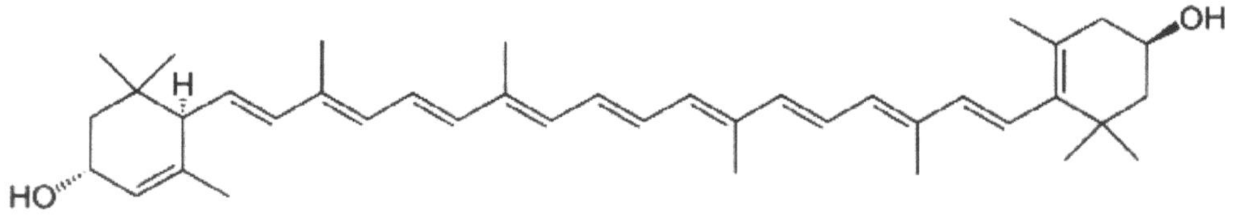

FIGURE 7.12   Lutein.

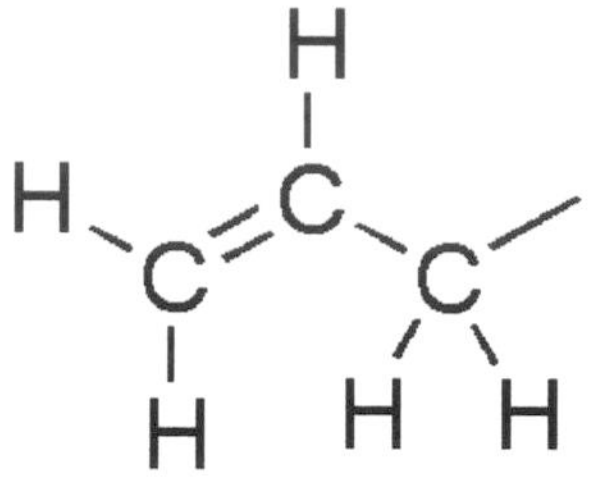

FIGURE 7.13   The allyl group.

The site on the saturated carbon atom is known as the 'allylic position'. A group attached at this site forms what may be referred to as an allylic compound. Thus, allyl alcohol, $CH_2 = CHCH_2OH$, has an allylic hydroxyl group. Since the C–H bonds in an allylic functional group are weaker than the C–H bonds located in ordinary tetrahedral carbon centers, such as in methane for example, allylic C–H bonds are more reactive. This heightened reactivity has many practical applications in organic chemistry.

Many allylic compounds have a lachrymatory effect on human beings (for more on allyl compounds, see also the chapter on 'Garlic and Pungent Smells' in Part IX.)

HUMAN HEALTH; VITAMIN A, RETINOL AND RETINAL

Vitamin A is the collective name for number of unsaturated nutritional organic compounds which includes retinol, retinal and several other carotenoids among which beta-carotene is the most significant. Vitamin A has multiple functions in that it is important for

- growth and development
- maintenance of the immune system and
- good vision.

β-Carotene is cleaved by the action of an enzyme in the human body to form retinol.

Retinol functions as a form of storage for the vitamin in the body. Chemical reduction of the alcohol functional group readily occurs in the body to give the visually active and corresponding aldehyde, retinal. Vitamin A is needed by the retina of the eye in the form of retinal, which influences low-light and color vision.

You will appreciate from Figure 7.10 that retinol is a diterpenoid compound. It has two isoprene units in the middle of the molecule. Many different geometric isomers of retinol and retinal arise from the *trans* or *cis* configurations in four of the five double bonds found in the polyene chain. Some *cis* isomers carry out essential functions in visual acuity. The 11-*cis*-retinal isomer, for example, is the chromophore of rhodopsin—the photoreceptor molecule in vertebrate animals. Vision depends upon light-induced isomerization of the chromophore from the *cis* to the *trans* isomer. This change of conformation activates the photoreceptor molecule. Night blindness, an inability to see well in dim light, is one of the earliest signs of vitamin A deficiency followed by decreased visual acuity. In 1967, George Wald won the Nobel Prize for his work with retina pigments, which led to the understanding of the role of vitamin A in vision.

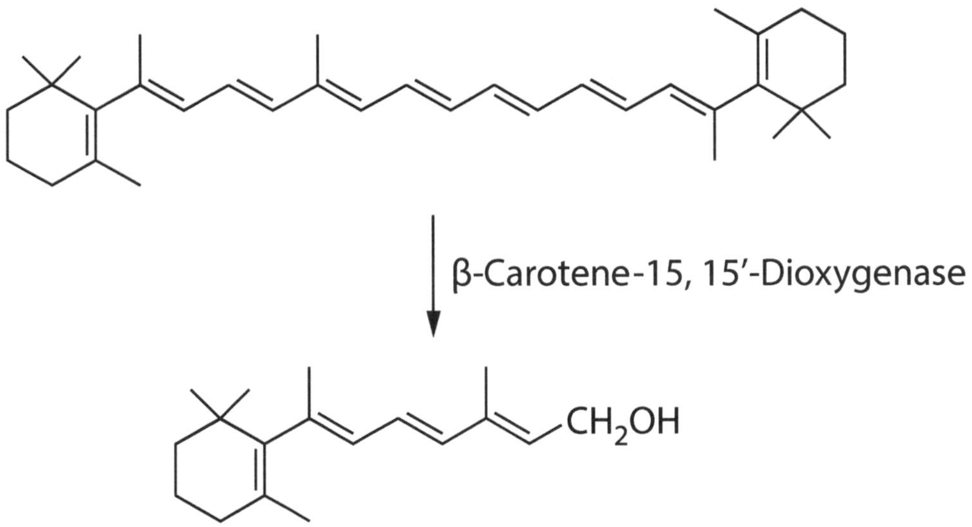

FIGURE 7.14 Conversion of β-carotene to retinol.

Source: Ray Cooper's book Natural Products Chemistry. Sources, Separations, and Structures (Fig. 11.2 p. 126) Taylor and Francis

Questions

1. The allyl functional group is very reactive. Illustrate the breadth of usefulness to organic chemistry of compounds containing the allyl group.
2. Explain why ultraviolet spectroscopy is a much more important tool for studying carotenoids than infra-red spectroscopy.
3. Explain the difference between the carotene carotenoids and the xanthophylls carotenoids and illustrate your answer with exemplar compounds.
4. Choose one tetraterpenoid and illustrate the nature and prevalence of geometric isomerism.
5. Explain why the addition of one or more functional groups containing a nitrogen atom or an oxygen atom to a molecule of a dye or pigment can change the color of the chromophore.

SUGGESTED FURTHER READING

K. Akhtari, K. Hassanzadeh, B. Fakhraei, N. Fakhraei, H. Hassanzadeh, and S. A. Zarei. 2013. *A Density Functional Theory Study of the Reactivity Descriptors and Antioxidant Behavior of Crocin*. Comput Theor Chem 1013: 123–129.

R. Cooper and G. Nicola. 2014. *Natural Products Chemistry; Sources, Separations and Structures*. CRC Press, Taylor and Francis Group.

D. B. Gregg. 1974. *Agricultural Systems of the World*, 1st Edition. Cambridge University Press.

T. Hill. 2004. *The Contemporary Encyclopaedia of Herbs and Spices; Seasonings for the Global Kitchen*. Wiley.

REFERENCE

A. Rubio-Morago, R. Catillo-Lopez, and L. Gomez-Gomez. 2009. *Saffron Is a Monomorphic Species Revealed by RAPD, ISSR and Microsatellite Analyses*. BMC Res Notes 2: 189.

WOAD (ISATIS TINCTORIA) AND INDIGO

Abstract: Indigo, a purplish-blue dye obtained from the plant Woad, is among the oldest of natural dyes used in textile dyeing, body art and printing.

Organic Chemistry

- *indigo—a purplish-blue dye*
- *the emergence of the modern synthetic dye industry*
- *the production and use of synthetic azo dyes.*

Context

- *indigo—a naturally occurring alkaloid*
- *indigo produced as a synthetic dye in the modern world*
- *indigo as a pigment in body painting and in the dyeing of textiles.*

INDIGO THROUGH THE AGES

Records indicate that even as late as the 1980s in countries of the southern Arabian Peninsula, people applied indigo as a healing treatment to their skin. They found that wrapping an indigo cloth coated in beeswax and oil around a wound was an effective antiseptic procedure.

Indigo became so valuable as a textile dye that it was traded extensively from East to West and ultimately drove the early development of the synthetic organic chemistry during the middle to late 1860s in Europe, through the leadership of such scientific giants as William H. Perkin and Adolf von Baeyer. While there are plenty of historical accounts of the use of indigo, many of which list countless ailments that indigo has been reputed to cure, it is the pigment and coloring properties of the alkaloid that are examined here.

HISTORICAL NOTE

Indigo is among the oldest of dyes used for textile dyeing and printing. In many Asian countries, (such as India, China, Japan, and South eastern Asia), indigo has been used as a dye over thousands of years. The dye was also known to ancient civilizations in Mesopotamia, Egypt, Britain, ancient Persia and Africa. As a highly valuable commodity, dyes from India spread west along established ancient trade routes. It was via Arab merchants that indigo came to the attention of the Roman Empire where it was considered a luxury item. Although in the 1st century AD the Roman historian, Pliny the Elder, refers in texts to the common use of indigo and woad by the Gallic tribes of northern Europe, indigo had been established in use since the Bronze Age (2,600–800 BC) particularly in Celtic Britain. Indigo was known as woad since it was obtained from the native wild plant known as Woad, Isatis tinctoria.

When the sea route from Europe to India was opened in the late 15th century, direct trade flourished since dangerous land routes and tax duties imposed by the Persians and other intermediaries were avoided. Ease of import of indigo through ports in Portugal, the Netherlands and England meant that use of the dye grew significantly in Europe.

Rising European demand for indigo stimulated the creation of indigo plantations in other parts of the world. Indigo became a major crop in both Jamaica and in South Carolina where enslaved Africans and African Americans met the need for labor. Indigo developed into a major export crop in the plantations of colonial South Carolina, which tended to reinforce dependence on slavery during the 18th century.

Owing to its high value as a trading commodity, indigo was often referred to as blue gold.

INDIGO FROM PLANTS

Most of the indigo obtained from natural sources came from plants in the genus *Indigofera*, which are native to the tropics. The primary commercial species in Asia is *Indigofera tinctoria*.

In Celtic Britain, where the dye was used as a kind of war paint by native tribes, indigo was known as woad since it was obtained from the native wild plant Woad or *Isatis tinctoria*.

The molecular structure of indigotin, commonly known as indigo, is shown in Figure 7.15.

Indigotin is directly related to the important, natural building block, indole, shown in Figure 7.16. Indole has been described in various chapters throughout the book. In particular, see also the chapters on 'Africa's Gift to the World' in Part II and 'Maca from the high Andes of South America' in Part IV.

Indole is relatively stable chemically, due to a high degree of electron delocalization above and below the plane of the molecule. In turn, the chemical stability of the indole building block markedly influences the chemistry of indigo, which is remarkably unreactive.

There is an interesting feature of indigo which requires some explanation. The molecule is planar, so the possibility of geometric isomerism has to be considered yet only the *trans* isomer has ever been isolated. This phenomenon is understood in terms of steric hindrance, especially between the neighboring oxygen atoms, which destabilizes the *cis* isomer. More on geometric isomerism is presented in the chapters on 'Saving the Pacific Yew Tree' in Part II and 'Cannabis and Marijuana' in Part V.

EXTRACTION OF INDIGO

The precursor molecule to the chemical, indigo, is called indican. Indican is a colorless organic compound, soluble in water, which occurs naturally in *Indigofera* plants. Indican is obtained by processing leaves of the plant, which contain a small percentage by weight of the compound. Indican is a glycoside (see also the chapter dealing with polysaccharides, 'Asian Staple—Rice' in Part III). In fact, the chromophore in many dyes and pigments is often chemically bound to a glucose moiety through a glycoside bond. As a glycoside, indican will undergo hydrolysis to yield glucose and a molecule known as indoxyl in which the chromophore is located. Upon exposure to air, indoxyl is readily oxidized to indigo.

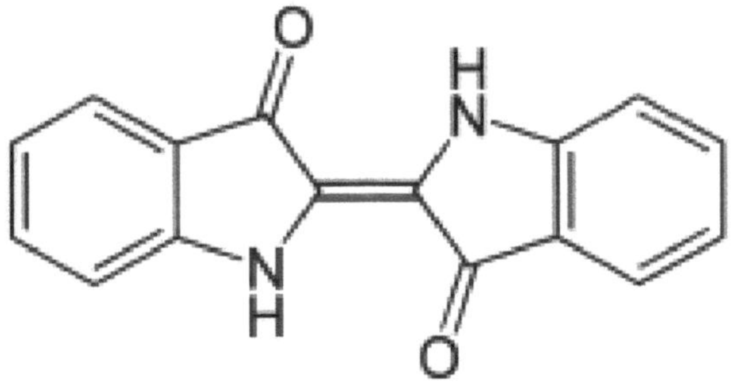

FIGURE 7.15   The structure of indigo, an alkaloid dye.

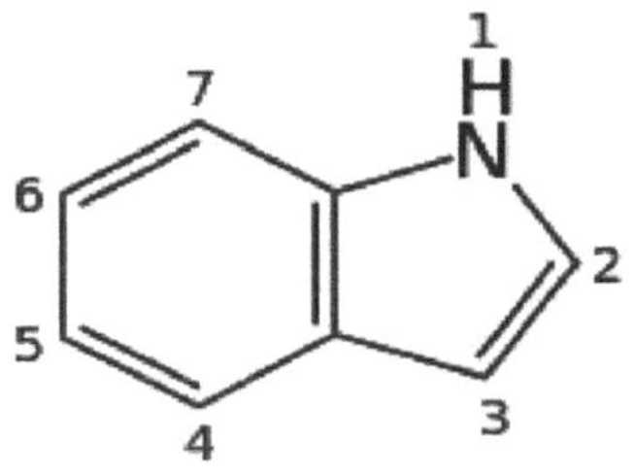

FIGURE 7.16   The structure of the natural building block, indole.

Thus, the extraction process involves soaking the leaves in water in the presence of air to release the blue dye as a precipitate. This material is then ground up with alkali and pressed into cakes, from whence it is dried and powdered. This powder can be mixed with various other substances to produce different shades of blue and purple.

Indigo is a dark blue crystalline powder which sublimes at 390–392°C. It is insoluble in many organic liquids such as water, alcohol and ether. The very low solubility of indigo in organic solvents is attributable to both intra-molecular and inter-molecular hydrogen bonding.

The molecule absorbs light in the orange part of the visible spectrum (λ_{max} = 613 nm) and there-fore the compound appears to the eye in the complementary color, purplish-blue. The chromophore (see Glossary) in indigo has been shown to be the central part of the molecule without influence from either of the two phenyl groups. Similarly to other conjugated compounds, the deep color arises from the delocalization of electrons above and below the series of three adjacent double bonds within the planar molecule. In this regard, it is not surprising that indigo and some of its derivatives possess the properties of organic semiconductors when deposited in thin films.

INDIGO AND DYEING TEXTILES

When indigo first became widely available in Europe in the 16th century, it was found that indigo was difficult to manage as a textile dye because it is not soluble in water. In order to overcome this difficulty, indigo was subjected to chemical reduction involving the nitrogen to oxygen double bonds. The reduction product is commonly known as indigo white or leuco indigo (Figure 7.18).

FIGURE 7.17   Indigo from natural sources.

Source: the historical collection of dyes at the Technical University of Dresden, Germany. Credit: Shisha-Tom. Creative Commons Attribution Share Alike 3.00 unported license Physical and Chemical Properties of Indigo

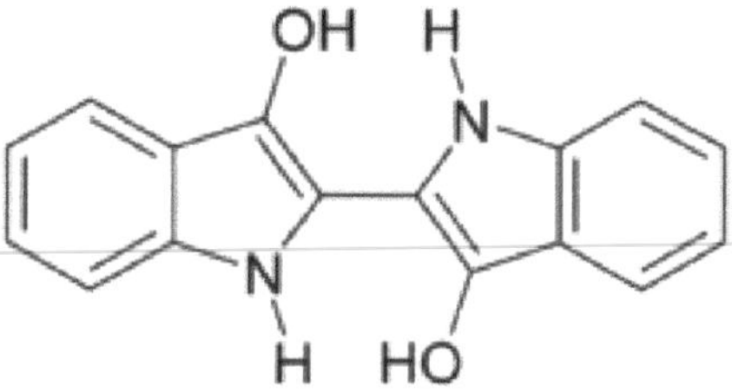

FIGURE 7.18   Indigo white or leuco indigo.

The process is as follows: when a submerged fabric made from cellulose is removed from the dye bath, white indigo quickly combines with oxygen from the air and reverts to the insoluble, intensely colored indigo, which is lodged in the weft and warp of the cellulose material by hydrogen bonding, since the cellulose polymer has many hydroxyl bonds (see cellulose in the chapter on 'Global Aloe' in Part VI and polysaccharides in the chapter on 'Asian Staple—Rice' in Part III). Relatively weak adherence through hydrogen bonding explains why the color is not fast, that is, it fades slowly, especially under the influence of regular washing of the fabric.

Levi Strauss and Wrangler Jeans

Indigo is primarily used as a dye for the cotton yarn used in the production of denim cloth in blue jeans. On average, a pair of blue jean trousers requires 3–12 g of indigo. According to Steingruber writing in *Ullmann's Encyclopedia of Industrial Chemistry*, about 20 million kg of indigo are produced annually and applied mainly in the dyeing of blue jeans. Small amounts are used for dyeing wool and silk. Synthetically produced indigo is used today for the dyeing of modern blue jeans made by such well-known companies as Levi Strauss and Wrangler.

Indigo is also permitted for use as a food colorant by the Food and Drugs Agency in the USA.

Lincoln Green Worn by Robin Hood

In medieval England, Lincoln Green was a favorite dyestuff for coloring woolen fabric. Lincoln Green was celebrated in the legendary story of the nobleman-turned-fugitive, Robin Hood, who operated from Sherwood Forest in Nottinghamshire to do good for many disadvantaged serfs and villains of the time. Garments colored in Lincoln Green would certainly have provided good camouflage within the forest for the band of followers led by Robin Hood.

The color Lincoln Green was formed from purple-blue indigo when blended with a yellow dye from the flowering wayside and garden plant, Tansy, *Tanacetum vulgare*.

HISTORICAL NOTE ON TANSY

Tansy has a long history of use. The ancient Greeks cultivated tansy for its use as an insect repellent. In modern times, tansy has been planted in potato fields to discourage the predatory Colorado beetle. Dried tansy has also been burned as incense on account of the camphor it contains. However, soaking the leaves and flowers in water and boiling the solution produces, on cooling, a mustard-yellow extract effective in dyeing wool.

The cyclic monoterpenoids thujone, myrtenol and camphor may be extracted in a volatile oil from the yellow flowers of this perennial, herbaceous plant of the aster family. These cyclic monoterpenoids all have a molecular formula of $C_{10}H_{16}O$. For more on cyclic terpenoids, see the chapters on 'Frankincense and Myrrh' and 'A Plant from the East Indies, Camphor' in Part VI and 'Saffron and Carotenoids' in Part VII.

A HISTORICAL NOTE ON THE MODERN SYNTHETIC DYE INDUSTRY REPLACING NATURAL DYES AT THE START OF THE 20TH CENTURY

Aromatic amines can be readily oxidized to a variety of compounds that possess intense color. The most famous of these synthetic dyes was discovered by accident in 1856 by the English chemist William H. Perkin, who was trying to find a means of synthesizing quinine while studying at the Royal College of Chemistry in London. Perkin's first experiments yielded only a black solid mess, which is often the result of a failedorganic synthesis. But how poor had his experimental technique really been? Perkin noticed that the black products were consistently reproducible. When cleaning his reaction flask with alcohol, the keen-eyed Perkin noticed and then saw the significance of the purple-mauve solution he had produced. Perkin obtained a patent from the British Patent Office for this colored material, which he named mauveine, having also found that the substance was effective in dyeing silk and other textiles. Shortly afterwards, he set up a factory to produce the dye, which became a commercial success.

Beyond the profound significance of this discovery as a technological breakthrough were the simple facts that (a) he used aniline as the starting material, which is readily available from coal and (b) the chemistry of the process is straightforward.

Perkin had found that when aniline is exposed to a strong oxidizing agent, e.g., acidified potassium dichromate solution, mauveine was produced along with a number of other oxidation products such as p-benzoquinone (or 1–4 benzoquinone), a bright yellow solid.

However, the molecular structure of mauveine proved difficult to determine and in fact was only finally identified in 1994 by Meth-Cohn and Smith (Meth-Cohn, et al., 1994). One type of mauveine molecule is known as mauveine A. It has the molecular formula, $C_{26}H_{23}N_4^+X^-$ and incorporates two building blocks of aniline within a fused five-ring structure.

There was a drive to reduce dependence upon natural sources of indigo. Advances in synthetic organic color chemistry from the time of William Perkin, continued by 1865 in the studies of synthetic dyes by the famous German chemist Adolph von Baeyer. He researched the synthesis of indigo and determined its chemical structure in 1870 and first described a synthetic method for producing indigo in 1878. However, the process was difficult in practice and still depended on plant extracts as source material, therefore a search for alternative chemical routes continued at the chemical laboratories of the German companies Badische Anilin und Soda Fabrik (BASF) and Hoechst. Eventually, an alternative process was identified, which had the potential to permit synthetic production of indigo on an industrial scale, starting from readily available aniline. By 1897, BASF had developed a commercial manufacturing plant. Von Baeyer was awarded the Nobel Prize for chemistry in 1905 partly in recognition for his work on indigo. Today, modern production of indigo is based on naphthalene as the feedstock.

Within just a few years of their discovery, synthetic dyes had almost completely replaced natural dyes because they

- offered a kaleidoscopic range of color and shade
- possessed a brilliant, intense presence
- made available colors which would act on a wide range of fabrics
- did not wash out easily
- did not fade on exposure to natural light
- presented opportunities to develop reliable color reproduction.

Nevertheless, human curiosity ensures, even well into the 21st century, such that interest in the potential of and experimentation with natural dyes remains undimmed.

Azo Dyes

In the modern world, the most common synthetic coloring agents used in textiles, in pharmaceutical products, in foods and in cosmetics are known as azo compounds.

Kekule, working in Germany in 1866, was the first chemist to identify the azo functional group, which comprises of the double-bonded nitrogen atoms as shown in Figure 7.19.

The azo group may be bonded to aromatic groups, which, therefore, indicate that these molecules have extended conjugated systems that act as chromophores and impart vibrant and dense color, usually red, orange or yellow. An example is shown in Figure 7.20. Electron delocalization also imparts stability to aromatic azo compounds.

One of the earliest commercial textile dyes produced was called chrysoidine, a yellow compound, which has been in use for dyeing wool since 1875. It was synthesized from aniline and m-phenylenediamine. The synthesis of azo dyes named Congo Red and Bismarck Black soon followed in the late 1800s.

Azo dyes can be made in the laboratory in a three-step process. Firstly, phenylamine is added to nitrous acid in the presence of hydrochloric acid, which produces benzenediazonium chloride. Sodium phenoxide is then produced separately by dissolving phenol in sodium hydroxide solution. When the two solutions are added together at a low temperature just above the freezing point of water, the benzenediazonium ion, an electrophile, combines with the phenate ion, a nucleophile, to yield a yellow precipitate—the azo compound. This reaction is shown in Figure 7.21.

In addition to dyeing textiles, azo dyes are also utilized as pigments in paints and as acid/alkali indicators in chemical analysis in the laboratory. Methyl red, for example, described in the chapter on 'Reversible Colors in Flowers, Berries and Fruit' in Part VII, is an azo dye.

A widespread application of azo dyes is as a colorant in the food processing industry. In the European Union, many azo food colorants are listed for brevity under what is known as an E number. For example, the azo dye, tartrazine, a well-known food colorant, is identified as E102. Tartrazine imparts a yellow color to a wide range of processed foods such as potato chips, corn chips, ice cream and breakfast cereals. Tartrazine is accepted for use in this way by the FDA. However, it should be noted that, in recent years, concern has grown among health professionals and nutritionists over the use of artificial additives in processed foods, including colorants, because of the possibility of chemical change in poorly understood metabolic pathways that may produce unwanted toxins or carcinogens.

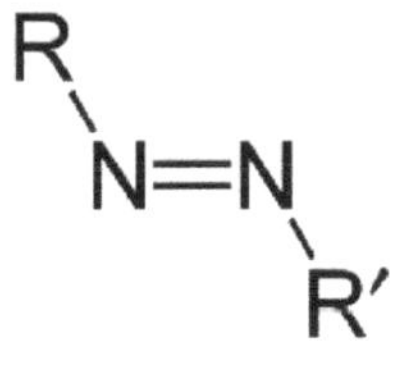

FIGURE 7.19   The generalized formula of an azo compound.

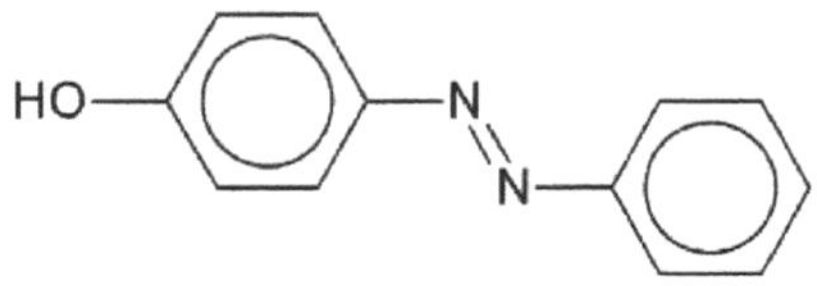

FIGURE 7.20   The molecular structure of an aromatic yellow azo dye.

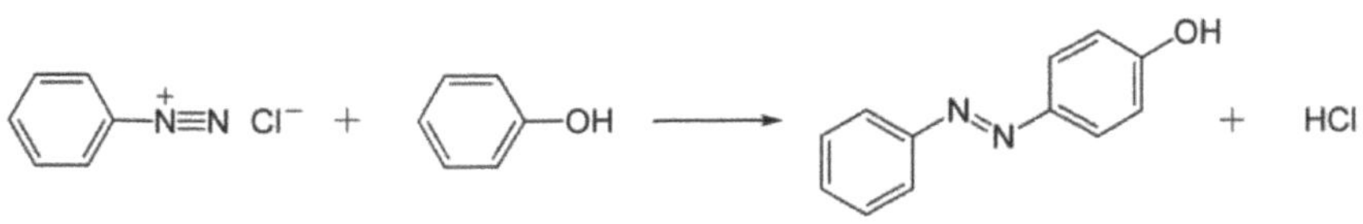

FIGURE 7.21   Coupling of benzenediazonium chloride and phenol.

Questions

1. Explain the phenomena of intra and inter molecular hydrogen bonding and show how they are responsible for the low solubility of indigo in organic liquids.
2. Explain the modern interpretation of the deep color and semi-conducting properties of indigo.
3. Although indigo white is closely related to indigo, explain why indigo white does not have any color.
4. Account fully for the low chemical reactivity of indigo.
5. Explain why adding functional groups to the chromophores of azo dyes may alter their color.
6. Draw the structure of the theoretical, geometric cis isomer of indigo

SUGGESTED FURTHER READING

J. Balfour-Paul. 1998. *Indigo*, pp. 218–220. British Museum Press.

J. Cannon and M. Cannon. 2007. *Dye Plants and Dyeing*. A & C Black.

Editorial Committee. 1964. *Dye Plants and Dyeing—A Handbook*. Brooklyn Botanical Garden.

A. Feeser. 2013. *Red, White and Black Make Blue: Indigo in the Fabric of Colonial South Carolina Life.* University of Georgia Press.

I. Grae. 1974. *Nature's Colors, Dyes from Plants*. Macmillan Publishers.

A. Kumar Samanta and A. Konar. 2011. 'Dyeing of Textiles with Natural Dyes', in *Natural Dyes*, edited by Perrin Akcakoca Kumbasar, pp. 29–56. InTech.

S. Robinson. 1969. *A History of Dyed Textiles*. Studio Vista Limited.

E. Steingruber. 2004. 'Indigo and Indigo Colorants', in *Ullmann's Encyclopedia of Industrial Chemistry*. Wiley-VCH.

REFERENCE

O. Meth-Cohn and M. Smith. 1994. *What Did W. H. Perkin Actually Make When He Oxidised Aniline to Obtain Mauveine?* J Chem Soc Perkin 1: 5–7.

RED DYES FROM HENNA, DYER'S BUGLOSS AND MADDER

Abstract: Many natural and artificial coloring substances, whether they are dyes or pigments, are derivatives of quinones which may be extracted from a range of plants. Quinones are second only to synthetic azo dyes in importance as dyestuffs.

Alizarin, the red coloring matter in paint well known to artists, was extracted originally from the Madder plant. It was the first natural dye to be synthesized from coal tar.

Organic Chemistry

- *quinones as building blocks in nature*
- *quinones as chromophores*
- *applications of quinones as pigments in paints and cosmetics and as dyes in food processing and textiles*
- *dyeing textiles*
- *color fastness, resistance to fading by ultraviolet radiation and repeated washing,*
- *mordants.*

Context

- *quinones; benzo, naphtha and anthra*
- *Lawsone, a red-orange dye from Henna—a naphthoquinone*
- *Alkannin, a red-brown dye from Dyer's Bugloss—a naphthoquinone*
- *Alizarin, a crimson red dye from Madder root—an anthraquinone*
- *Parietin, an orange dye from Lichens—an anthraquinone*
- *substances in use for millennia in various cultures as hair dye, in body art and tattoos and in the modern world as cosmetics and food colorants.*

QUINONES AS BUILDING BLOCKS IN NATURE

Quinones are oxidized derivatives of aromatic compounds: benzene and those with fused rings, naphthalene and anthracene.

However, it is important to note that while quinones are conjugated molecules they are not aromatic. As we have seen with other organic compounds, conjugation of unsaturated groups within a molecule does influence greatly the color chemistry of quinones. The chromophore is the quinone group, which is why the color of many quinones, whether benzoquinones, napthoquinones or anthraquinones, is red to orange in hue. The concept of chromophores is dealt with in greater depth in the chapter on 'Our World of Green Plants—Human Survival' in Part VII.

Many natural and artificial coloring substances (dyes and pigments) are derivatives of quinones. They are second only to azo dyes in importance as dyestuffs. Alizarin (1,2-dihydroxy-9, 10-anthraquin one), the red coloring matter extracted from the Madder plant, was the first natural dye to be synthesized from coal tar. The name, Madder, comes from the term in Arabic, al'isara, meaning 'pressed juice'.

   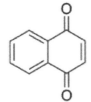   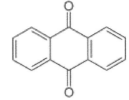

1,4-Benzoquinone   1,4-Naphthoquinone   9,10-Anthraquinone

FIGURE 7.22   Representative quinones.

FIGURE 7.23   Dyeing fabrics in Morocco.

Source: file released under the Creative Commons Attribution-Share Alike 2.5 Generic license

NAPHTHOQUINONES

Lawsone, a Red-Orange Dye from Henna

An extract from the plant Henna (*Lawsonia inermis*) has been used as a colorant for over three thousand years and is often mixed into a paste and used to color skin and hair. Henna has been used as a cosmetic dye on hair and skin and as the pigment in tattoos for millennia especially on the Indian sub-continent. Lawsone combines chemically with the protein known as keratin in hair and skin resulting in a strong permanent stain that lasts until the skin or hair is shed. Lawsone strongly absorbs ultraviolet radiation and in aqueous solution is an effective sunscreen.

Even today, henna is used in Ayurvedic medicine for the treatments of rheumatism, insect bites, skin ailments, burns and wounds to name a few. It is also proven to have antifungal and antibacterial properties that are linked to the active component, lawsone—the same chemical that is responsible for its color and properties as a dye.

You can see clearly why the naphthoquinones take their systematic name from the building block, naphthalene.

Alkannin, a Red-Brown Dye from Dyer's Bugloss

Dyer's Burgloss, *Alkanna tinctoria*, is a member of the borage family of plants and provides alkannin, a red-brown dye used in food coloring and in the production of cosmetics, especially lipstick. Alkannin is a naphthoquinone with a pentene side chain.

ANTHRAQUINONES

The anthraquinones are a common family of naturally occurring substances with yellow, orange and red pigmentation. Anthraquinoids are a class of compounds based upon the anthraquinoid skeleton (see Figure 7.22).

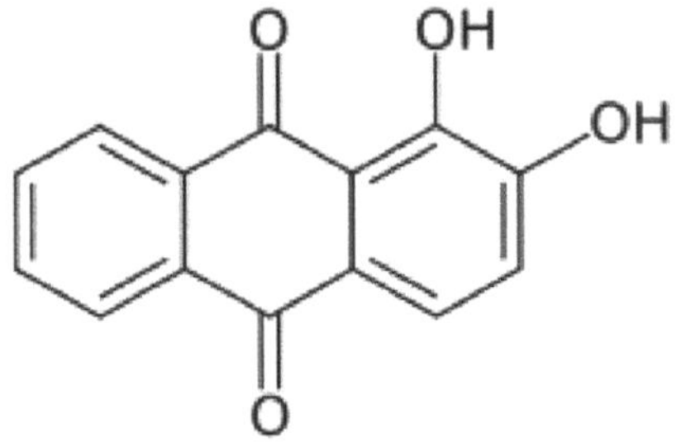

FIGURE 7.24 Lawsone or 2-hydroxy-1,4-naphthoquinone.

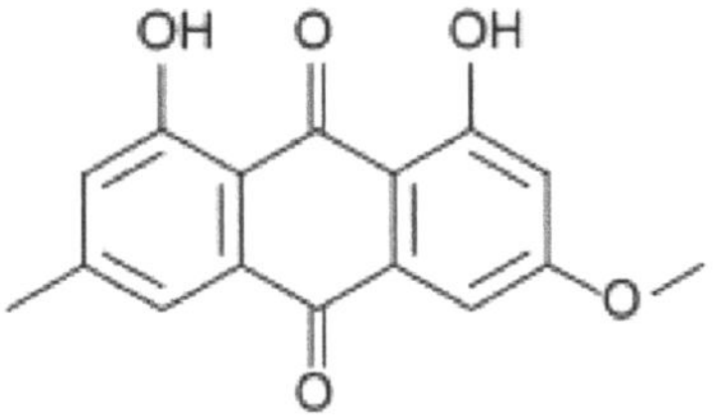

FIGURE 7.25 Alkannin.

Plant-derived medicinal laxatives containing anthraquinones were presented earlier in the book in the chapter on 'Global Aloe' in Part VI. The two key compounds aloin (Figure 6.20) and emodin (Figure 6.21) have been used as traditional medicines since antiquity.

Yet, many anthraquinones, including emodin, also have a long history of use as dyes (Bien et al., 2005). Emodin dyes wool fibers yellow and polyamide fibers red. Fabrics made from these threads show high uptake of this natural anthraquinone and so emodin has significant potential as a useful alternative to synthetic dyes. There are many natural and artificial coloring substances (dyes and pigments) that are derivatives of quinones. They are second only to azo dyes in importance as dyestuffs.

Alizarin Red from Madder Root

Early evidence of the use of plant extracts for dyeing textiles comes from Sindh in modern-day Pakistan where cotton dyed with a red substance from the plant Madder, *Rubia tinctorum*, has been recovered from archaeological sites dating from 3,000 BC. Alizarin is the crimson red dye involved.

Madder was widely used as a dye in Western Europe too in the late medieval period. In 17th-century England, alizarin was the red dye used for the uniforms of the New Model Army led by the parliamentarian Oliver Cromwell during the English Civil War, which began in 1642. The distinctive red color would continue to be worn by soldiers in subsequent centuries giving the English Army, later the British Army, the nickname 'redcoats'.

Alizarin was one of the first natural dyes to be produced synthetically in 1869, which led to the collapse of the market of the dye from its natural source.

Alizarin is the main ingredient for the manufacture of pigments known to painters as Rose Madder and Alizarin Crimson. A notable use of alizarin in modern times occurs in biochemical assay to determine, quantitatively by colorimetry, calcium and calcium compounds, which are stained red or light purple (Puchtler et al., 1969). Alizarin also continues to be used commercially as a red textile dye.

Parietin from the Lichen, Xanthoria Parietina

Lichens are plant organisms involving fungi and algae living in symbiosis (see the chapter on 'Our World of Green Plants' in Part VII). Lichens can grow almost everywhere in the world including hot desert climates and snow packed environments but favor moist, temperate zones where they can be found on rocks, roofs, walls and tree trunks.

FIGURE 7.26   Alizarin or 1,2 dihydroxy anthraquinone.

Lichens have been used to provide textile dyes, especially for wool, over many, many centuries in northern climes. No mordant is needed to set the color. Many lichens that were used to dye cloth were fermented with ammonia to produce the dye. The natural source of ammonia was urine!

During the Middle Ages, one species of lichen, *Rocella*, was extremely important and known as Royal Purple and as an extract was used exclusively by the church and royalty for purple robes.

One commonly found lichen is Crottal or Crottle, *Parmelia saxatilis*, which furnishes a reddish brown dye. It is also known as the skull lichen because it is light gray color and form.

Parietin is an orange dye extracted from the lichen, *Xanthoria parietina*, which is the orange-yellow lichen found in abundance on rocks, walls and tree trunks. The lichen usually grows in exposed positions, and as parietin absorbs in the blue end of the spectrum it protects the lichen from sunlight.

Parietin is the natural dye used for centuries to color both Scottish tartan and the famous cloth woven and dyed in the Outer Hebrides of Scotland called Harris Tweed (Dweck, 2002).

By the early 20th century, natural dyes had been entirely replaced by organic synthetic dyes as supplies of these were readily available, and demand for Harris Tweed was growing rapidly worldwide. In the modern world, tartan and Harris Tweed have literally become established as the stuff of modern fashion icons!

Although in modern times lichens are not much used as sources for the dye industry, they do provide an important role in monitoring air pollution. Because most lichens are sensitive to air pollution, the health of the lichen species can be mapped and monitored and used as a proxy measure of air quality especially in industrial zones.

Incidentally, you can see that parietin is a little unusual as a naturally occurring compound in having an ether linkage, a relatively unreactive functional group (see also the chapter on 'Maca from the High Andes in South America' in Part IV.)

Textile Dyes—Color Fastness and Mordants

Solubility of Dyes in Water

In order to permeate the fibers of the fabric, dyes need to be soluble in water, ideally because water is a cheap and readily available solvent that is straightforward to use. Water is a polar solvent that will readily dissolve polar or ionic solutes due to hydrogen bonding. Sometimes, the molecules of a dye are fully covalent and need to be made polar in order to dissolve the dye in water. This may be done for instance by incorporating a sulfate ion in the molecule of a dye to produce a sodium or potassium salt.

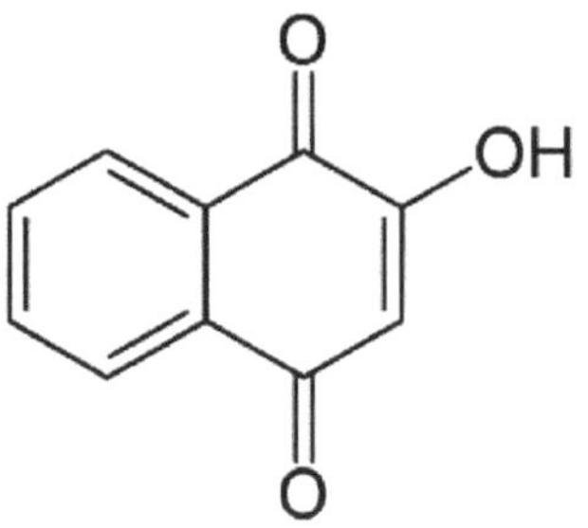

FIGURE 7.27   The common lichen, Xanthoria parietina.

Source: www.freebigpictures.com

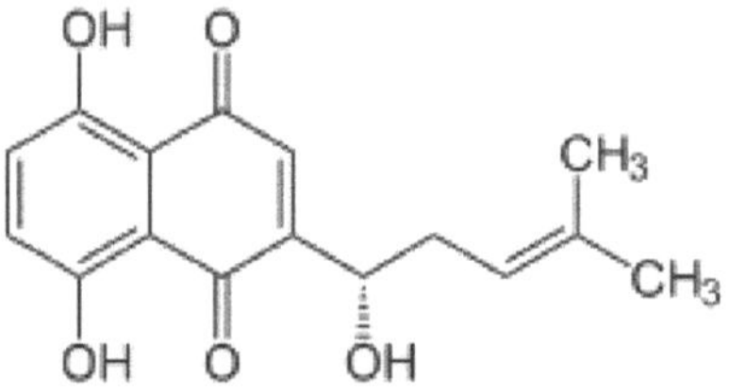

FIGURE 7.28   Parietin.

Color Fastness

Dyes have to be colorfast to be effective, meaning that over time the dye should neither fade nor wash out of the fabric easily. For this to be so, physical or chemical interactions have to take place between the dye and the fabric.

The fibers of naturally occurring fabrics such as cotton and linen are composed of cellulose, which has many hydroxyl groups along its length. Should the dye molecule contain the amine functional group, linkage of the fiber and dye by the electrostatic forces of hydrogen bonding is perfectly feasible. The problem is that such an arrangement does not lead to permanent color dyeing of the textile.

The stronger electrostatic forces of ionic bonding can be taken advantage in some circumstances. A case in point is where the reverse to the aforementioned applies, that is, the dye molecule contains an hydroxyl group as part of a carboxylic acid function, while the fiber of the fabric has a complementary alkaline component in the form of an amine group. The acid functional group donates a proton to the alkaline amine thus creating a distribution of negative and positive centers leading to ionic bonding between the dye and fiber. This situation applies in the dyeing of the natural fabrics of silk and wool and to the artificial fabric, nylon. Color fastness due to ionic bonding is improved compared to that available through weaker hydrogen bonding.

Of course, the ultimate in color fastness is achieved through covalent chemical binding of the molecules of dye to the polymers of the fiber. Thus, retention of permanent color by the fabric is the result of chemical reaction between a suitable functional group on the dye molecule and either the hydroxyl group or the amine group typically present in the polymeric molecules of fibers.

Mordants

A mordant is a substance used to affix a dye to a fabric (or to a tissue section in preparation for analysis in the biology laboratory). A particular mordant is chosen specifically for each combination of dye and fabric.

The mordant is usually an aqueous solution of one of a variety of metal salts; those of chromium, aluminum, tin, copper or iron. An example of a material used as a mordant is the mineral deposit, alum, which is a source of chromium or aluminum ions.

In practice, the fibers of the fabric are first treated with an aqueous solution of the metallic salt and then the dye is introduced. The metal ions are involved in the formation of strong chelates (or co-ordination complexes) which bind the fiber and the dye.

For background on chelates, see also the chapter on 'Our World of Green Plants—Human Survival' in Part VII. Established interpretation of the formation of chelates is in terms of Ligand Field Theory, which stresses covalent bonding between the ligands and the central ion—and Crystal Field Theory—which emphasizes ionic interactions. These theories account for the perturbing of the outer orbitals occupied by the valency electrons of the central enclosed ion to create quantum steps between energy levels that correspond to the energy available at frequencies in the visible spectrum of light. Hence, absorption of light causes color in many chelates too. Full explanations of Ligand Field Theory and Crystal Field theory are given in books, for instance by J. N. Murrell et al. and by G. M. Barrow, which are mentioned under the sub-heading concerning suggestions for further reading.

Questions

1. Alizarin is one of ten isomers of dihydroxyanthraquinone. How many can you identify?
2. There are only two isomers of benzoquinone. Give their systematic names. Knowing the organic chemistry of the alkene and ketone functional groups, describe the kinds of chemical behavior you would expect of these conjugated molecules.
3. Account for the fact that quinone molecules are strong chromophores and give examples of the commercial use of these substances as dyes and pigments.
4. Explain differences between dyes that are direct in operation and those dyes that require fixing with a mordant.
5. Lichens are particularly sensitive to the presence of compounds of sulfur. Explain how studies of lichens over time can yield information about the distribution and intensity of air pollution.
6. Certain molecules in organic chemistry, especially those containing the oxygen atom or the nitrogen atom, can act as ligands to form chelates or co-ordination compounds with the ions of metals such as magnesium and iron. Explain why the color of copper sulphate changes dramatically from white, as an anhydrous ionic salt, to blue in aqueous solution and then to dark blue when ammonia or an amine such as methyl amine is added to the solution.
7. What is a mordant? Comment on the fact that Bixin, a yellow dye (see the chapter on 'Saffron and carotenoids—yellow and orange dyes'), does not require the use of a mordant when it is applied in the dyeing of fabrics that are based on cellulose such as cotton.
8. Phthalo Blue is a dark blue pigment used in paints and ink that was developed synthetically in the 1930s. Phathalo Blue has the empirical formula $C_{32}H_{16}N_8Cu$. Explain how the central copper atom forms a co-ordination compound with phthalocyanine, which has four nitrogen atoms, each with a lone pair of electrons. Also, explain what is meant by a colloidal suspension of Phthalo Blue in water since this is commonly known as ink.

SUGGESTED FURTHER READING

G. M. Barrow. 1966. *Physical Chemistry*. McGraw-Hill.
J. N. Murrell, S. F. A. Kettle, and J. M. Tedder. 1965. *Valence Theory*. John Wiley.

REFERENCES

H.-S. Bien, J. Stanwitz, and K. Windelich. 2005. *Anthraquinone Dyes and Intermediates*. Ullmann's Encyclopaedia of Industrial Chemistry.
A. C. Dweck. 2002. *Natural Ingredients for Colouring and Styling*. Int J Cosmet Sci 24 (5): 287–302.
H. Puchtler, S. Meloan, and M. Terry. 1969. *On the History and Mechanism of Alizarin Red Stains for Calcium*. J Histochem Cytochem 17 (2): 110–124.

REVERSIBLE COLORS IN FLOWERS, BERRIES AND FRUIT

Abstract: The brilliant colors of red, purple and blue in many naturally occurring and cultivated ornamental flowering plants—and their fruit and berries—are due to the production of a class of organic compounds called flavonoids of which the anthocyanins are an important group.

Organic chemistry

- *color in flowers and cross pollination*
- *color in berries and fruit and food and seed dispersal*
- *the reversible colors of the acid/base indicators used in chemistry and biology*
- *the significance of phytochrome—a photo reversible pigment.*

Context

- *flavonoid dyes and pigments and*
- *anthocyanins*
- *photoperiodism and phototropism in plants.*

THE INESTIMABLE VALUE OF COLOR IN FLOWERS, BERRIES AND FRUIT

The red, blue and purple colors of many, many naturally occurring and cultivated ornamental flowering plants is due to the production of a class of organic compound called anthocyanins in the flowers and other tissues such as leaves and stems. Particularly clear examples of such flowering plants are pansies (Figure 7. 29), violets and snapdragons while red cabbage and rhubarb are representatives of plants with colored leaves and stems.

A great variety of insects is attracted to colorful flowers and in many instances there is a specific relationship between the flowering plant and a given insect. The outcome is cross pollination of the flower and regeneration and dispersal of the species.

Color in berries and fruit is also due to the presence of anthocyanins. While most plants produce anthocyanins, certain plants are particularly rich in them, notably blueberry, cranberry, raspberry, blackberry, blackcurrant, cherry, grapes, red-fleshed peaches and tomatoes.

These strong colors attract herbivores and omnivores from the animal kingdom. As a consequence, the food provided results in the deposition of indigestible seeds and the dispersal of the plant species whose growth is also conveniently given a boost of fertilizer in the process!

FIGURE 7.29 Purple pigmentation due to anthocyanins in the common pansy.

Source: Rosendahl. Public domain

FIGURE 7.30 Strawberries, raspberries, blackberries and blueberries.

Source: www. wallpaperup.com

THE BEAUTIFUL COLOR OF AUTUMN LEAVES

In the autumn, the red color of leaves in deciduous plants, notably trees, is due to the presence of anthocyanins, which according to Archetti et al. are actively produced toward the end of summer (Archetti, et al., 2011). The yellow and orange colors of autumn leaves are the result of the breakdown of chlorophyll which exposes the carotenoids present, xanthophylls (yellow) and carotenes (orange)—see the chapter on 'Saffron and carotenoids—yellow and orange dyes' in Part VII.

FLAVONOIDS

Several phenolic classes have been presented in earlier chapters, for example the lignans and lignin from a phenylpropanoid skeleton. This chapter introduces an important class of phenolic natural products, namely the flavonoids. Flavonoids are the largest class of polyphenols. Chemically, they may be defined as a group of polyphenolic compounds consisting of substances that have two substituted benzene rings connected by the chain of three carbon atoms and an oxygen bridge, as shown in Figure 7.31.

Flavonoids are ubiquitous in plants and products obtained from them such as fruit, vegetables, herbs and beverages such as tea and red wine. Flavonoids are associated with a broad spectrum of beneficial effects on human health attributable to their antioxidative, anti-inflammatory, anti-mutagenic and anti-carcinogenic properties.

The three rings labeled A, B and C are arranged in a fixed pattern. One position in the central ring is occupied by an oxygen atom rather than by a carbon atom. Each atom in the ring is identified by being given a number, running from 1 through 8 for the adjacent rings A and C, while ranging from 1′ through 6′ on the offset ring B.

FLAVONOIDS AND ANTHOCYANINS

The flavonoid class of compounds contains the C6-C3-C6 skeleton. Based on the degree of oxidation of the 3-C bridge and whether the bridge is open or closed in a fused ring, flavonoids can then be subdivided into six major groups, among which are the flavonols, flavones, isoflavones and anthocyanins (Figure 7.32).

ANTHOCYANINS—REVERSIBLE DYES

Anthocyanins are chromophores (Figure 7.33), which absorb in the green to yellow parts of the visible spectrum of light dependent upon the pH of the aqueous solution, in which the anthocyanin is

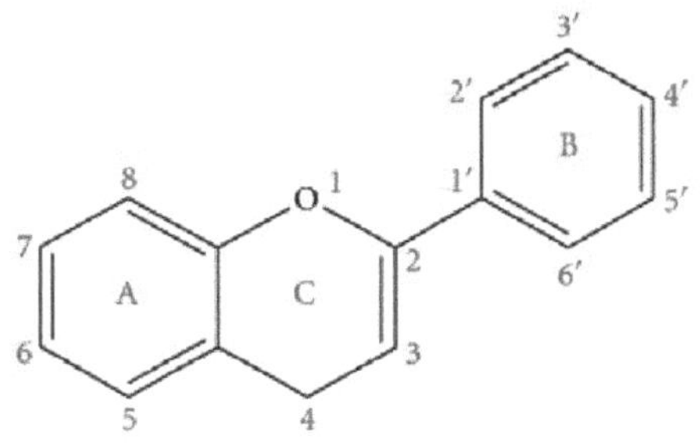

FIGURE 7.31   Common flavonoid carbon skeleton structure.

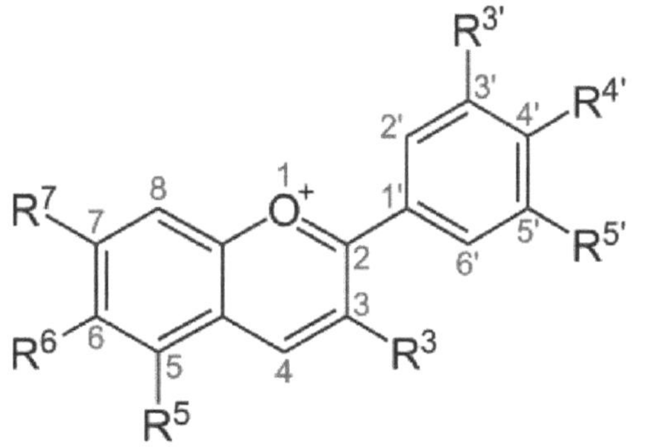

FIGURE 7.32   The generalized molecular structure of anthocyanins where R represents different functional groups.

dissolved, thereby producing the red to purple to blue coloration in light reflected from or transmitted through the tissues of the plant.

Anthocyanins may also act as a sunscreen for the plant as they absorb in a different part of the visible spectrum to chlorophyll.

Anthocyanins are secondary metabolites that find application as a food colorant and are approved for use as such in the European Union, Australia and New Zealand (UK Food Standards Agency, 2023).

Since anthocyanins are water-soluble substances, they can easily be extracted with water from a range of colored tissues in plants such as leaves (red cabbage); flowers (petals of the pelargonium, poppy, rose); berries (blueberries, blackcurrants) and stems (rhubarb). Anthocyanins are red in acidic solutions, purple in neutral conditions, blue in basic solutions and colorless in strongly alkaline solutions.

As these color changes are quite reversible, anthocyanins extracted from plants can provide a crude pH indicator.

REVERSIBLE DYES AND ACID/BASE INDICATORS

Litmus

Litmus is a naturally occurring, water-soluble mixture of different dyes extracted from lichens, especially *Roccella tinctoria* (see also the chapter entitled 'Our world of green plants—Human Survival' in Part VII). Litmus is a naturally occurring pH indicator. The name arises from Norse meaning literally 'colored moss'. The color of litmus changes from red in acidic solution to blue in alkaline solution. The term litmus test has entered everyday language as a metaphor for a test that purports to distinguish authoritatively between alternatives.

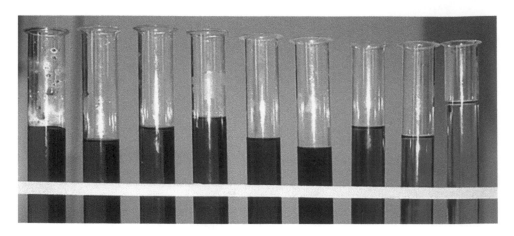

FIGURE 7.33   Anthocyanins from red cabbage change color reversibly dependent on the pH of the solution, from left to right 1, 3, 5, 7, 8, 9, 10, 11 and 13.

Source: Permission is granted to copy, distribute and/or modify this document under the terms of the GNU Free Documentation License from Indikator-Blaukraut. JPG: Supermartl

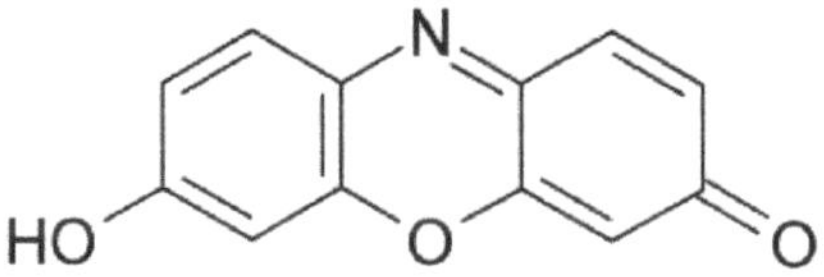

FIGURE 7.34   Phenoxazone.

Litmus compounds belong to a family of substances known as phenoxazones (Figure 7.34), which have complex structures involving a common, unsaturated core involving nitrogen and oxygen in fused rings. The unsaturated core is the chromophore.

As with other indicators, litmus is a weak acid that partially dissociates in water into a hydrogen ion and an anion. The position of the equilibrium is affected when acid or alkali is added to the solution. Acid causes the equilibrium to be displaced toward the un-dissociated molecule, which is red, while alkali shifts the equilibrium toward the anion as hydrogen ions are removed. The anion is blue. Neutral solution occurs at a pH of around 7, but the purple color is reached rather slowly and indistinctly, which explains why litmus is often used simply to indicate whether a solution of a compound is either acidic or alkaline. The position of the equilibrium is readily reversible caused by the addition or removal of hydrogen ions.

Methyl Red and Methyl Orange

Methyl red is a reversible azo dye, which is made synthetically as was explained in the chapter on 'Woad and Indigo'. Indicators, such as methyl red and methyl orange, are usually weak acids or bases. Methyl red is used in microbiology to identify bacteria, which produce acids from the fermentation of glucose whilst methyl orange is employed in chemistry as an indicator in titrations of weak bases against strong acids. A core structure is common to both substances, which relates to a common reversible color change from red through orange to yellow. The molecular structure of methyl red is shown in Figure 7.35.

Methyl red is a weak acid, which will partially dissociate in water at the functional group of the carboxylic acid. In acidified solution, a hydrogen ion from the strong acid attaches to one of the double-bonded nitrogen atoms thereby breaking the double bond leaving only a single bond

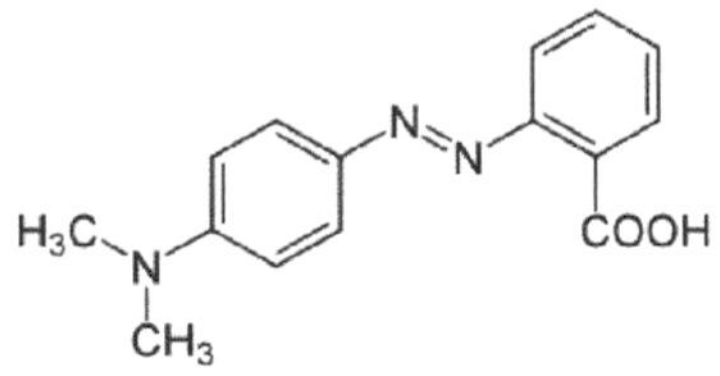

FIGURE 7.35 The molecular structure of methyl red.

between the neighboring nitrogen atoms. Both the un-dissociated molecule and the anion are structurally modified and the positive charge from the hydrogen ion is distributed over the structure. As a different chromophore is now present the solution appears red because the chromophore absorbs in the blue/green area of the visible spectrum.

The process is reversible. In alkaline solution, the anion appears yellow because it absorbs in the blue area of the visible spectrum. The yellow form of the anion reappears because the nitrogen atom is not protonated and a different chromophore is present with a structure similar to that shown in Figure 7.35.

While the color change is reasonably sharp, methyl red turns orange at a pH of about 4. An equal mixture of the two different anions in methyl orange produces an orange color at about pH 5 rather than at the neutral value of pH 7 so methyl orange is only useful as an indicator in certain circumstances, namely in titrations of weak bases against strong acids.

Phenolphthalein

Phenolphthalein is another indicator used in acid-base titrations. Phenolphthalein does not occur naturally—it is a synthetic compound. Phenolphthalein is also weak acid which dissociates into an H^+ ion and a large anion in solution. Phenolphthalein molecules are colorless in acidic solutions and pink in alkaline solutions when the phenolphthalein anion is predominantly present. When a base is added to the phenolphthalein solution, the molecule/ion equilibrium shifts to the right, as H^+ ions are removed, leading to the release of more anions.

PHYTOCHROME—A REVERSIBLE PIGMENT AND A BIOLOGICAL LIGHT SWITCH

The pigment phytochrome brings deciduous green plants 'back to life' in the spring in temperate zones of the world and also helps them to set their daily rhythm. Phytochrome consists of a molecule of bilin within which there is an open, linear chain of four building blocks of pyrrole (see Figure 7.38).

Bilin may be compared with the closed rings containing four cyclic pyrrole molecules that are found in chlorin and heme covered in the chapter on 'Our World of Green Plants—Human Survival' in Part VII. You will immediately appreciate that phytochrome is not a flavonoid (and therefore not an anthocyanin) but it does have very interesting and hugely significant properties as a chromophore.

Many flowering plants and deciduous trees make use of phytochrome to assess the relative length of night and day to set their circadian rhythm (or twenty-four-hour cycle). It is, perhaps, the most important factor (temperature being another) in sensing the changing of the seasons in temperate climes. The circadian rhythm is so named from Latin, *circa* meaning *about* and *dies* the word for a *day*. Thus, a plant can set a period for flowering and a deciduous tree can grow fresh leaves for the spring season and shed leaves in the autumn season.

Phytochrome is a photo-reversible pigment. Phytochrome is sensitive to red light, which is readily available from sunlight during the day, absorbing strongly at a wavelength of 650 nm (Figure 7.36).

When phytochrome (shown as PR in the absorption spectrum) absorbs energy, it changes physically (shown as PIR) to adopt the form of another conformer (see Glossary and Figure 7.37).

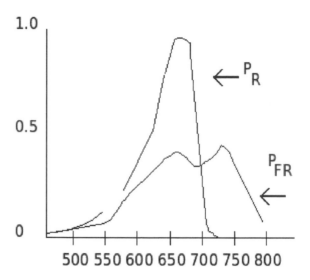

FIGURE 7.36 Phytochrome absorption spectrum in the cereal oats.

Source: Audriusa. Creative Commons Attribution Share-Alike 3.0 Unported license

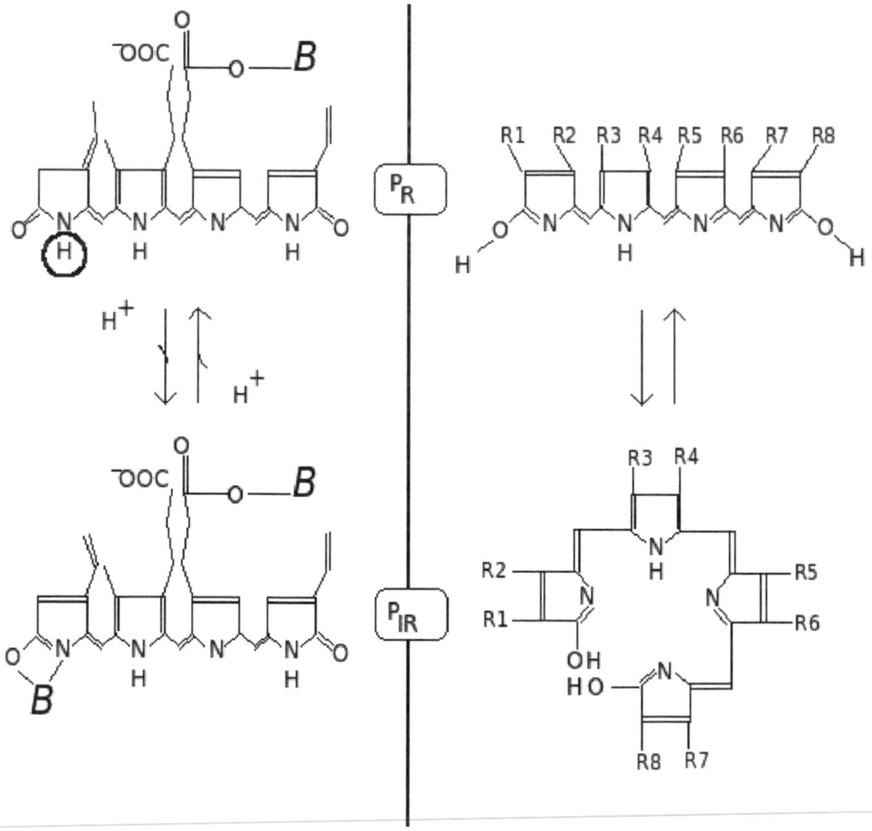

FIGURE 7.37 Two ideas that try to account for the change of shape in photochrome (PR—red form, PIR—infra-red form).

Source: Audriusa. Creative Commons Attribution Share-Alike 3.0 Unported license

The conformer PIR is, in turn, able to absorb strongly at a lower frequency and longer wavelength of 725 nm, which lies just within the infra-red region. When the conformer absorbs, it reverts to the ground state (PR) once again. Infra-red radiation is present at night, when the environment cools and there is no red light from insolation, and so the ground state conformer becomes plentiful in the plant once again by dawn.

PHYTOCHROME IN PHOTOPERIODISM AND IN PHOTOTROPISM

Phytochrome FR is biologically active in the plant whereas Phytochrome R is not.

Many green plants use this reversible change from one conformer of phytochrome to another as both a biological switch and as a means of monitoring the length of darkness in relation to the length of daylight. Hence, plants are able to regulate both their circadian and seasonal rhythms.

Phytochrome also enables plants to grow toward light—a property known as phototropism.

Questions

1. Compare and contrast the properties of the indicators litmus, methyl red, methyl orange, thymol blue and phenolphthalein and explain the most suitable applications for each one in titrations.
2. Explain what is meant by the term end point or equivalence point in a titration. Why are indicators not used for titrations of weak acids against weak alkalis?
3. Distinguish between the major sub divisions of the flavonoids; flavanols, flavones, isoflavones and anthocyanins.
4. Resveratrol is a flavonoid with an open bridge structure found in the skin of grapes.

   Explain why the compound has a chromophore and is responsible for the color of red wine.
5. Give a full account of the significance to green plants—and in turn to the Earth as whole—of the reversible photo properties of the pigment, phytochrome.

SUGGESTED FURTHER READING

J. Buckingham and V. R. N. Munasinghe. 2015. *Dictionary of Flavonoids*. CRC Press, Taylor and Francis R. N. Munasinghe.
J. B. Harborne. 1964. *Biochemistry of Phenolic Compounds*. Academic Press.
L. C. Sage. 1992. *A Pigment of the Imagination, a History of Phytochrome Research*. Academic Press.
W. Vermerris and R. Nicholson. 2006. *Biochemistry of Phenolic Compounds*. Springer.

REFERENCES

M. Archetti et al. 2011. *Unravelling the Evolution of Autumn Colors: an Interdisciplinary Approach*. Trends Ecol Evol 24 (3): 166–173.
UK Food Standards Agency. 2023. *Current Approved Additives and Their E Numbers*. UK Food Standards Agency.

Part VIII

Plant Materials

INTRODUCTION

Plant materials have been employed ubiquitously in all human cultures and in a variety of forms from raw state to the heavily processed and from utilitarian to decorative. Applications are many and varied: housing construction, furniture, basketry, cordage, lashings, textiles, writing materials, personal accessories, transport, musical instruments, artwork and tools—in short, in most aspects of human life. Selection of plant material is dependent upon inherent physical properties of tensile strength, flexibility, shape, density or surface morphology and/or myriad of chemical properties. A summary of the chemistry in Part VIII follows.

Chapter	Organic Chemistry	Context
Wood	carbohydrates,	shelter, fuel, tools, art
	non sugars	climate mitigation
	cellulose and starch	
Rubber	polymerization	human dependence
	allyl functional group	commercial significance
	vulcanization	
	physical properties, molecular structure	
	colloids and emulsions	
Bioplastic	polymerization	biodegradability
	carboxylic acids, alcohols, esters	environmental issues
	sugars	sustainable practice
	condensation reactions	
	chirality	

DOI: 10.1201/9781032664927-8

THE WONDERFUL WORLD OF WOOD

Abstract: *Since time immemorial, most human cultures have depended upon trees and their wood: for heating and cooking, for provision of shelter, for simple everyday objects such as tools and dug-out canoes, for functional furniture and a multitude of artifacts of aesthetic value. Despite the introduction of many new materials in modern times, trees retain their value and physical appeal yet have a new, significant role to play in helping to mitigate the threat of climate change.*

Two naturally abundant polysaccharides, cellulose and starch, are found in trees, and the associated organic chemistry is considered.

Organic chemistry

- *carbohydrates, the polysaccharides*
- *non sugars; cellulose, starch and inulin*
- *lignin*

Context

- *human exploitation for shelter, fuel, tools and aesthetic value*
- *environmental significance of trees as a sink for CO_2.*

WHAT IS WOOD?

Wood is the structural tissue found in the trunk, branches and roots of trees. It is composed of cellulose fibers, which are strong in tension, embedded in a matrix of lignin that resists compression (Figure 8.1).

Organic compounds comprise 99% of the total mass of dry wood; 49% being carbon, 44% oxygen, 6% hydrogen and 0.1–0.3% nitrogen.

Applications

Wood has been used for thousands of years as a construction material (Figure 8.2), for fuel, for tools and weapons (Figure 8.3), for furniture (Figure 8.4) and for producing paper. More recently it has emerged as a feedstock for the production of purified cellulose and its derivatives, such as cellophane and cellulose acetate. As an abundant, sustainable resource, wood remains of value as a source of renewable energy.

FIGURE 8.1   Symbol of strength and rigidity—an aged oak tree.

FIGURE 8.2   Building construction, a hammer-beam roof.

Source: Laurel Fan. Creative Commons Attribution-Share Alike 2.0 Generic license

FIGURE 8.3   An Australian boomerang—a hunting tool.

Source: Adrian Barnett. Public Domain

CARBOHYDRATES THAT ARE NON-SUGARS

Cellulose belongs to a class of organic compounds called carbohydrates, which are the most abundant found in living organisms. The generic name, carbohydrate (*carbon hydrates*), arises from the observation that the molecular formula of this class of compounds is $C_n(H_2O)_n$, where *n* is typically a large number. Carbohydrates are produced by photosynthesis: the condensation of carbon dioxide requiring energy in the form of light and the green pigment, chlorophyll (see the chapter on 'Our World of Green Plants—Human Survival' in Chapter VII for more details).

$$nCO_2 + nH_2O + energy \rightarrow C_nH_{2n}O_n + nO_2$$

FIGURE 8.4 Aesthetic appeal, tilt-top table and marquetry.

Source: Richard Walter. Creative Commons Attribution 2.5 Generic license

Carbohydrates may be sub-divided into *sugars* and *non-sugars* and examples of each are presented in Table 8.1. Carbohydrates that are sugars are presented in detail in *'Asian Staple—Rice'* in Part III.

While this very large class of compounds is usually referred to as the carbohydrates in organic chemistry, the same class of compounds is typically known as the saccharides in biochemistry, whereas in food science the term relates rather loosely to food that is particularly rich in carbohydrates.

Sugars, given the suffix, *ose*, can be monosaccharides, disaccharides or oligosaccharides. The term saccharide is derived from the Greek word for sugar, 'sacchar'. Sugars are colorless, crystalline solids that are soluble in water and usually have a sweet taste.

In contrast, *non-sugars* are white, amorphous solids insoluble in water. Non-sugars can be broken down by chemical hydrolysis into their constituent sugars, which is why non-sugars are classed as polysaccharides. Polysaccharides are very large polymeric molecules indeed and can attain extremely high molecular weights of over 100,000 Da (see Glossary).

If hydrolysis breaks down a non-sugar into only one type of sugar, the non-sugar is known as a *homo-polysaccharide*, whereas if more than one different sugar is released the non-sugar is described as a *hetero-polysaccharide*. Polysaccharides are widely distributed in nature. *Cellulose* and *starch* are probably the most abundant organic compounds known.

Cellulose is a good example of a homo-polysaccharide and is represented by the formula $(C_6H_{10}O_5)_n$ where *n* ranges from 500 to 5,000 depending on the natural source of the polymer. Cellulose is found in wood (50%), jute (70%), hemp and flax (80%) and in the seed hairs of the cotton plant (almost 100%). On account of hydrogen bonding between chains of cellulose along their whole length, it is difficult for molecules of a solvent to intrude. As a consequence, cellulose is insoluble in most solvents.

Starch is also a homo-polysaccharide. While cellulose is completely insoluble in water, starch is only very partially soluble in water. Due to its extremely high molecular weight and to inter-polymer hydrogen bonding mentioned earlier, starch forms a colloidal suspension when mixed with water.

TABLE 8.1

Sugars and Non-Sugars Are Carbohydrates

Carbohydrates	
Sugars	Non-Sugars
Monosaccharides	*Polysaccharides*
Glucose	Starch
Fructose	Cellulose
Oligosaccharides	
Sucrose	
Lactose	

Starch may be hydrolyzed under suitable conditions to glucose. Starch is the substance in which plants store their reserves of carbohydrates and is typically found in bulbs, tubers and seeds. The main commercial sources of starch are rice, wheat, maize and potatoes.

Another homo-polysaccharide is *inulin*—also represented by the generic formula $(C_6H_{10}O_5)_n$. Inulin is obtained from the tuber of the dahlia and yields fructose when hydrolyzed in the laboratory. Inulin is a huge linear polymer where n can lie between 70 and 200 fructose units. The dahlia comes to mind as an ornamental plant originating in Mexico, where it is regarded as a national emblem. Dahlia tubers, however, are a source of a little protein, some digestible carbohydrate and a lot of polysaccharide fiber due to the presence of inulin, which is in indigestible by bacteria resident in the human alimentary canal. Most of the inulin is excreted as fiber. Unsurprisingly, human consumption of dahlia tubers soon disappeared once the Aztec empire had fallen to the invading Spanish conquistadors.

Galactomannans are an example of a hetero-polysaccharide made up of the sugars galactose and mannose within the polymer—hence the name. Galactomannans may be added to processed food products such as ice cream to increase viscosity and hence improve texture (for more on cellulose and galactomannan visit the chapter on 'Global Aloe' in Chapter VI).

Monosaccharides and Polysaccharides Are Inter-Changeable in Plants

In plant cells, monosaccharides, sugars, are the primary source of energy driving metabolism and biosynthesis. When monosaccharides are not immediately needed by cells, they are converted into polysaccharides. In plants, starch is the polysaccharide used for storage of surplus monosaccharides and for conversion back into energy. However, the most abundant polysaccharide is cellulose, which is the structural component of the cell walls of plants and many forms of algae.

THE VALUE OF CELLULOSE

Shelter and Culture

Mankind has exploited the physical properties and beauty of cellulose in the form of wood from trees from the earliest times, particularly for shelter, for practical items such as furniture and basic tools and to create objects d'art. In more recent times, other applications have emerged.

Commercial Applications

Cellophane is a thin, transparent sheet made from regenerated *cellulose*. Cellophane has low permeability to air, oils, bacteria and water, which makes it very useful for food packaging. Cellophane is totally biodegradable and is effectively a polymer of glucose similar to cellulose. Cellulose film has been manufactured continuously since the mid-1930s and is still used today. In the UK and in many other countries, 'cellophane' is a registered trademark. However, in the USA the term is often used informally and more generally to refer to a wide variety of plastic wrapping films that are not composed of cellulose.

Explosives

When cellulose is immersed in a mixture of nitric and sulfuric acids, the hydroxyl functional groups of cellulose are replaced by the by the nitrate group with the elimination of water. As the reaction involves an alcohol and an acid it is an example of esterification represented simply by the equation

$$cellulose + acid = cellulose\ ester + water.$$

When up to six of the hydroxyl functional groups are replaced on each glucose unit, cellulose nitrate may be explosively ignited in air and is known commonly as gun cotton. The explosion produces a low pressure shock wave. Cellulose nitrate (or nitro cellulose, as it is commonly known) is a constituent of cordite, which was the material used to propel projectiles, bullets or shells until it was superseded by other products after the end of World War II.

Textiles

In industry, cellulose can be processed into fine threads which have been used to produce different artificial fibers known commercially as 'Rayon', 'Celanese' and 'Tricel'. Artificial fibers of cellulose are also woven into fabrics with natural fibers such as cotton or wool to lengthen the durability of garments made from them.

Historically, cloth has been woven from cellulose fibers obtained from diverse plant sources, prominent among them are cotton and flax for linen. Sailcloth was also prepared from cotton, for lightness and linen, for strength, although these natural fibers rotted easily in warm, humid conditions at sea (Figure 8.5).

Bindings

Again based on cellulose, bindings such as ropes and lashings have been prepared from fibrous material extracted from hemp, sisal and the dried stems and leaf ribs of the banana plant. In the days of sailing ships, these sources were of vital significance although today ropes for maritime purposes are made mainly from artificial fibers since they are much more durable and water resistant.

THE VALUE OF STARCH

Human Diet

As an indispensable item of the human diet, starch is encountered daily in potatoes, bread, rice and cakes.

Brewing of Beer

Cereal grain is the main commercial source of *starch*. Starch from barley grain is used in the brewing industry. The process involves a number of steps.

Initially, the grain is kept in moist, warm conditions. Starch is converted to the sugar or disaccharide, maltose, having the molecular formula $C_{12}H_{22}O_{11}$, by the enzyme, diastase, which is present in barley.

$$2(C_6H_{10}O_5)n + nH_2O = (C_{12}H_{22}O_{11})n$$

Water is applied to dissolve the maltose and yeast is added to the liquor. Yeast contains two enzymes, maltase and zymase, which bring about fermentation in the final two steps to make beer.

$$C_{12}H_{22}O_{11} + H_2O = 2C_6H_{12}O_6$$

FIGURE 8.5   The schooner, *Lewis R. French,* built in 1871.

Source: Raphodon. Creative Commons Attribution-Share Alike 3.0 Unported, 2.5 Generic, 2.0 Generic and 1.0 Generic license

Here, maltose is converted to the monosaccharide glucose by maltase.

$$C_6H_{12}O_6 = 2C_2H_5OH + 2CO_2$$

Finally, ethanol is generated from glucose through the action of zymase. For more on the process of fermentation, see the chapter on 'Chinese Cordyceps—Winter Worm, Summer Grass' presented in Part III.

Adhesives and Stiffening Agents in Fabrics

Starch is also widely employed in the preparation of adhesives and in stiffening agents for textiles and fabrics.

Bio-Plastic

As the cost of oil rises and the effects of global warming intensify, governments and industries are forced to turn attention to viable, natural alternatives. Bioplastics are made partly or wholly from materials derived from biological sources such as sugar cane, potato starch or cellulose from trees and straw.

Bioplastics are also often biodegradable and are, therefore, a much more sustainable product and enrich the soil on decomposition. Plant-based bioplastics can also be recycled. There is the additional benefit that biomass feedstock absorbs carbon dioxide as it grows thereby lowering the carbon footprint of the final product. Products and packaging made from bioplastics also have direct appeal to consumers (see the chapter on 'The Green Credentials of Bioplastic').

Starch is used to produce various bio-plastics, synthetic polymers that are biodegradable. An example is polylactic acid based on the glucose form of starch.

Lignin—the Structural Partner of Cellulose

Lignin is a very large, cross-linked, phenolic polymer forming a 3D structure that is both abundant and ubiquitous in the natural world since it is found in the cell walls of all green plants—especially those of trees.

Together, lignin and cellulose provide a vital structural function in plants, analogous to that of epoxy resin and glass fiber in a fiber-glass boat. Lignin provides rigidity while cellulose bears the physical load. Grasses, including cereals, have lignin content less than 20% by weight and so are pliable, easily bending under their own weight. However, trees are able to grow much taller and more rigidly due to additional lignin content of up to 30% by weight (Figure 8.6).

The formation of a polymeric matrix of lignin occurs through multiple combinations (through condensation reactions) of phenylpropanoid building blocks. Due to the presence of conjugated aromatic rings and carbonyl groups, lignin contains chromophores that absorb in the near UV spectrum (300–400 nm).

Phenylpropanoids derive their name from the two structural parts that make them up, namely, a six-carbon, aromatic phenyl group and a three-carbon tail of propene (see the chapter on 'Wheat—Ancient and Modern' for more on phenylpropanoids). Phenylpropanoids are a diverse family of organic compounds that are synthesized by plants from the amino acids phenylalanine and tyrosine. Phenylpropanoids are found throughout the plant kingdom, where they are essential components of

FIGURE 8.6   The tallest plant on Earth, redwood trees *(Sequoiadendron giganteum).*

a number of structural polymers, provide protection from ultraviolet light, defend against herbivores and pathogens and also furnish the floral pigments and scents that promote fertilization by attracting pollinators (see the chapter 'Colorful Chemistry, A Natural Palette of Plant Dyes and Pigments').

DISTRIBUTION OF TREES ACROSS THE WORLD

There are approximately 64,100 known species of tree in the world (Pappas, 2023) distributed over tropical, subtropical, temperate and cold temperate zones in many countries.

According to an estimate in 2015, the number of trees in the world is 3.04 trillion. The estimate also suggests that about 15 billion trees are cut down annually and about 5 billion trees are planted. In the 12,000 years since the start of human agriculture, the number of trees worldwide has decreased by 46% (Crowther et al., 2015 and Ehrenberg, 2015).

CONSERVATION AND ENVIRONMENTAL SIGNIFICANCE

Trees play a significant role in reducing erosion and moderating the climate. They remove carbon dioxide from the atmosphere and store large quantities of carbon in their tissues. In much of the world, forests are shrinking as trees are cleared to increase the amount of land available for agriculture. The Food and Agriculture Organization of the United Nations defines deforestation as the conversion of forest to other land uses (Figure 8.7). The FAO estimates that the global forest carbon stock decreased by 0.9% and tree cover by 4.2% between 1990 and 2020. Agricultural expansion continues to be the main driver of deforestation and forest fragmentation (FAO and UNEP, 2020).

Climate change is resulting in ongoing increase in global average temperature, which and produces attendant environmental consequences. Contemporary rise in global average temperature is

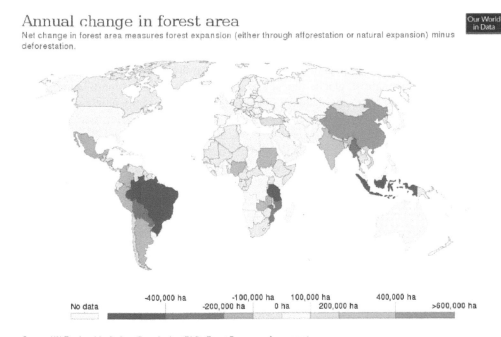

FIGURE 8.7 Annual change in forested area at December 2012.

Source: Hannah Ritchie and Max Roser. Creative Commons Attribution-Share Alike 4.0 International license

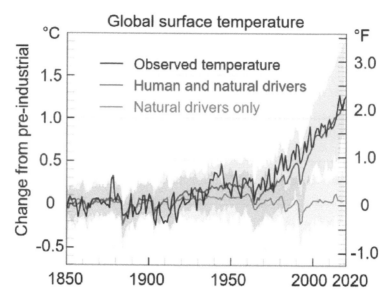

FIGURE 8.8 Global temperature change.

Source: Efbrazil. Creative Commons Attribution-Share Alike 4.0 International license

more rapid than any previous natural change as it is being caused primarily by humankind burning large volumes of fossil fuels since the industrial era began in the 1850s (Lynas et al., 2021) (Figure 8.8). Deforestation and some agricultural and industrial practices also influence the increase in greenhouse gases, notably carbon dioxide and methane. Greenhouse gases absorb some of the heat that the Earth radiates after it warms from sunlight. Larger amounts of these gases trap more heat in Earth's lower atmosphere, causing global warming.

Climate change is causing a range of increasing impacts on the environment. Deserts are expanding. Heat waves and wildfires are becoming more common. Warming in the Arctic and Antarctic has contributed to melting permafrost, glacial retreat and sea ice loss. There are more intense storms, droughts and other weather extremes. Climate change threatens human societies with increased flooding, food and water scarcity and economic loss. The World Health Organization (WHO) considers climate change to be the greatest threat to global health in the 21st century (WHO, 2015).

Reducing emissions necessitates generating electricity from low-carbon sources such as sustainable biomass rather than continuing to burn fossil fuels. Electricity generated from non-carbon-emitting sources will need to replace fossil fuels for powering transportation, heating buildings and operating industrial facilities (United Nations Environmental Program, 2019).

Significantly, plants can also make a contribution as carbon sinks for carbon can be removed from the atmosphere by increasing forest cover and by adopting farming practices that capture carbon in the soil (United Nations, Intergovernmental Panel on Climate Change (IPCC), 2019).

Questions

1. Explain what is meant by the term carbohydrate, and describe how this broad concept may be further classified into sub groups of compounds. Where do the following substances appear in this classification: glucose, sucrose, fructose, maltose, cellulose and starch? Show how polysaccharides fit into the picture, give an example of one and explain its use or value.

2. Give an account of the biochemical processes involved in the conversion of the polysaccharide cellulose, obtained from straw and wood chippings as by-products in agriculture and forestry, into the bio-fuel ethanol.

3. Give an example of a homo-polysaccharide containing three sugars and a hetero-polysaccharide also containing three sugars each with one to four linkages.

4. Where is lignin found in plants? Why is it important in plant growth?
5. Explain how re-forestation and tree species of rapid growth may make a contribution to climate mitigation.

REFERENCES

T. W. Crowther et al. 2015. *Mapping Tree Density at a Global Scale.* Nature (7568): 201–205. DOI: 10.1038/nature14967.

R. Ehrenberg. 2015. *Global Count Reaches 3 Trillion Trees.* Nature. DOI: 10.1038/nature.2015.18287.

FAO and UNEP. 2020. *The State of the World's Forests 2020. Forests, Biodiversity and People.* DOI: 10.4060/ca8985en.

IPCC. 2019. *Summary for Policymakers*, p. 18. IPCC.

M. Lynas et al. 2021. *Greater Than 99% Consensus on Human Caused Climate Change in the Peer-Reviewed Scientific Literature.* Environ Res Lett 16 (11): 114005. DOI: 10.1088/1748-9326/ac2966.

S. Pappas. 2023. *Thousands of Tree Species Remain Unknown to Science.* Sci Am 326 (5).

United Nations Environmental Programme. 2019, p. xxiii, Table ES.3; Teske, ed. 2019, p. xxvii, Fig.5 and Table ES.3 & p. 49; NREL 2017, pp. vi, 12

WHO. 2015. *IPCC AR5 SYR 2014*, pp. 13–16. WHO.

SECRETS OF A STICKY SECRETION

Abstract: The immense importance of natural rubber derives from its physical properties, foremost being elasticity although good electrical insulation and waterproofing are also significant. In England during 1770, the famous chemist Joseph Priestley observed that a piece of the material was extremely good for rubbing out pencil marks on paper and the name stuck.

While synthetic rubber has been produced since the 1940s, current production levels of natural and synthetic rubber are approximately equivalent and rising.

Organic chemistry

- *polymerization*
- *allyl functional group*
- *vulcanization*
- *the physical properties of rubber in relation to its molecular structure*
- *colloids and emulsions*

Context

- *human dependence on an astounding variety of products made from rubber*
- *commercial significance of rubber.*

LATEX

Deriving from the Latin word for liquid, latex is the term applied to a viscous fluid present in plants—not be confused with sap. It is a complex emulsion that coagulates on exposure to air and is exuded as tissue injury occurs. It serves mainly as defense against herbivorous insects.

Latex is produced by 20,000 flowering plant species from over 40 families of which about 12,000 plant species yield latex containing rubber, though in the vast majority of cases the rubber is not suitable for commercial use (Bowers, 1990).

However, the term latex is most strongly associated with the rubber tree (*Hevea braziliensis*).

NATURAL RUBBER

Almost all commercial production of natural rubber comes from plantations of the rubber tree, *Hevea braziliensis*, originally native to Brazil. The species grows well under cultivation. A properly managed tree will yield latex over several years. Cutting the bark of this tree releases the milky exudate, latex, which is an emulsion—an aqueous, colloidal suspension of large molecules of various hydrocarbons along with those of trace compounds such as proteins and fatty acids (Figure 8.9).

Latex is treated by evaporation or centrifuge in order to increase the concentration of hydrocarbons to about 70%. The task can also be achieved by introducing a dilute solution of acetic acid (ethanoic acid), which coagulates the colloidal droplets (Figure 8.10).

Cleaning removes contamination. Dried material is then stored or shipped abroad to be processed into various grades of rubber.

Natural rubber is used extensively in many applications and products, either alone or in combination with other materials. The most useful properties are those of high elasticity and resistance to water. By the end of the 19th century, industrial demand for rubber-like materials began to outstrip natural rubber supplies leading to the synthesis of synthetic rubber in 1909. In 2022, 15.4 million tons of synthetic rubber was produced. However, natural supplies of rubber remain hugely important. Between 1972 and 2022, global production of natural rubber rose from 3.06 to 14.2 million metric tons annually (Statista Research, 14/07/2023). Thailand, Indonesia, Malaysia, India and China are leading producers of natural rubber. Synthetic rubber is derived from petroleum feedstock and so the price of natural rubber is also determined, to a large extent, by the prevailing global price of crude oil.

Other potential sources of natural rubber have been examined and in some cases exploited. Many plants produce forms of latex rich in isoprene polymers but these sources entail elaborate processing

FIGURE 8.9   Latex being collected from a tapped rubber tree.

Source: Creative Commons Attribution-Share Alike 1.0 Generic license released into the public domain

FIGURE 8.10   Removing coagulated latex from troughs of dilute acid.

Source: Mitchell P. This file is licensed under the Creative Commons Attribution-Share Alike 3.0 Unported license

and fail to produce commercially viable quantities of rubber. Dandelion 'milk' contains latex. This latex exhibits the same quality as the latex from rubber trees but the latex content is low. In the 1940s in Nazi Germany during the hardship of the Second World War, research was directed toward the use of dandelions as a base for rubber production, but these efforts were not successful.

The latex of some plants is used for a specific purpose such as that of the chicle tree for chewing gum. In Central America, both the Aztecs and Maya chewed chicle gum to stave off hunger and provide a degree of oral hygiene. By the 1960s, most chewing gum companies had switched from using chicle gum to butadiene-based, synthetic rubber which is cheaper to manufacture and reliably available.

HISTORICAL INTRIGUE

Since early trade in natural rubber was heavily controlled by business interests in Brazil, the country remained the main source of latex rubber during much of the 19th century, although no laws expressly prohibited the export of seeds or plants. In 1876, Henry Wickham smuggled Amazonian rubber tree seeds out of Brazil and delivered them to Kew Gardens in the UK. Seedlings were propagated and sent onward to warmer climes in the British Empire; India, Ceylon (Sri Lanka), Singapore and Malaya (Malaysia). In Singapore and Malaya, commercial production was heavily promoted by Sir Henry Nicholas Ridley, who served as the first Scientific Director of the Singapore Botanic Gardens from 1888 to 1911 (Cornelius-Takahama, 2001). He distributed rubber seeds to many planters and developed the first technique for tapping trees for latex without causing serious harm to the tree (Leng and Keong, 2011). Before long Malaya became a leading supplier of natural rubber.

ORGANIC CHEMISTRY OF RUBBER

Rubber is a natural, elastic polymer of isoprene and is therefore unsurprisingly known as polyisoprene and as an elastomer. Isoprene, a building block in nature, is considered more fully in the chapter entitled 'Saving the Pacific Yew'.

In fact the isomer cis-1,4-polyisoprene is the main constituent of natural rubber (Figure 8.11). The polymer has a molecular weight of 100,000 to 1,000,000 daltons (see Glossary). Some natural rubber sources do produce a rubber called gutta-percha, which is composed of trans-1,4-polyisoprene, a structural isomer having similar properties.

Any one of the four possible isomers may be produced in synthetic polyisoprene. Isoprene, however, is expensive although buta-1,3-diene is readily available as a feedstock from natural gas and petroleum. A standard, general-purpose synthetic rubber is manufactured from the polymerization of buta-1,3-diene (3 parts) and styrene (1 part). A representative section of the polymer chain, which averages about 10,000 carbon atoms in length, has the sequence butadiene-butadiene-butadiene-styrene. Physical properties of synthetic rubber in terms of elasticity, density and durability are similar yet definitely different to those of natural rubber.

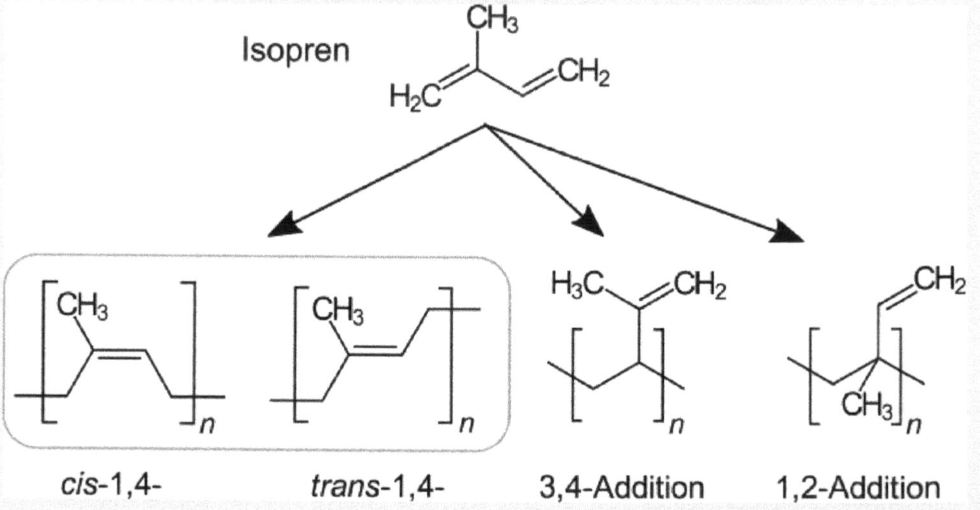

FIGURE 8.11   The isoprene monomer and the four isomers of polyisoprene where *n* is a large number.

Source: Roland chem. Creative Commons CC0 1.0 Universal Public Domain Dedication

The presence of the reactive, allyl functional group in each repeat unit of the polymer means that natural rubber is susceptible to oxidation in the atmosphere and is capable of vulcanization (for more on the allyl functional group see *Saffron and Carotenoids: Yellow and Orange Dyes*).

Crude rubber has some undesirable features; it softens and becomes sticky since it undergoes oxidation in the air and it has low tensile strength. A natural rubber polymer of polyisoprene has a long, zigzag shape in three dimensions that intermingles in random fashion with other neighboring rubber polymers. An elastomer is created by treating polyisoprene with just a few percent by weight of sulfur atoms, which creates extensive and durable cross-linking among polymers. The process is known as vulcanization.

Cross-linked rubber can be stretched by up to a factor of ten from its original length and, when released, it returns very nearly to its original length; an exercise that can be repeated many times with no apparent degradation in the rubber.

Although Mesoamericans had used stabilized rubber for balls and other objects as early as 1600 BC (Hosler, et al., 1999), Charles Goodyear established an industrial process in 1839 by researching the effect of heating crude rubber with sulfur, which resulted in rubber of greater elasticity and toughness (Slack, 2002). Vulcanized rubber was also found to be much less susceptible to oxidation, is insoluble in many solvents including water and is more stable over a wide range of temperature. During vulcanization, the long chains of isoprene in the polymer of rubber become cross-linked through covalent bonds formed with sulfur atoms, thus forming a three dimensional matrix. The optimal percentage of sulfur is found to be approximately 10%. In vulcanized rubber, polyisoprene molecules straighten when the rubber is stretched. Cross linkage makes the vulcanized rubber stronger and more rigid without loss of elasticity. Vulcanization of rubber creates di and polysulfide bonds between chains, which results in chains that tighten more quickly for a given strain, thereby making the rubber harder and less extensible.

Carbon black, derived from a petroleum refinery or a natural incineration processes, may be used as an additive to improve the strength of rubber—especially for vehicle tires—and to provide depth of color. Carbon black amounts to little more than amorphous particles of carbon soot in which there are short, unsaturated chains of carbon atoms in double and triple bonds. These unsaturated sites readily react with the unsaturated allyl functional groups in polyisoprene, thus providing strong cross-linkage between polymers; in this instance via carbon-carbon bonds rather than those between atoms of sulfur. When vulcanization and carbon black are both involved, rubber becomes hard—with a high coefficient of restitution—and durable, ideal for use in vehicle tires (Figure 8.12).

FIGURE 8.12   Modern Goodyear tires at an automobile race.

Source: Brian Cantoni. Creative Commons Attribution 2.0 Generic license

Charles Goodyear and Harvey Firestone were early pioneers and rivals in tire manufacture in the USA.

MYRIAD APPLICATIONS

Natural rubber offers good elasticity, while synthetic materials tend to offer better resistance to environmental stresses such as oil, temperature variations, chemicals and ultraviolet light. Uncured rubber is used in cements, for insulation in blankets and footwear (Figure 8.13) and in high-value products such as surgical wear. Of course, common everyday objects such as pencil erasers and elastic bands are fashioned from rubber.

FIGURE 8.13   Galoshes made by compression molding.

Source: public domain worldwide

Vulcanized rubber has many more applications. Resistance to abrasion makes harder rubbers valuable for the treads of vehicle tires and fan belts, conveyor belts and pump housings and piping. The flexibility of rubber is valuable in hoses, tires and the rollers of various types of machinery ranging from domestic items to printing presses. High elasticity makes rubber suitable for vehicle shock absorbers and for mountings designed to reduce vibration within machines. The high impermeability of rubber renders it suitable for air pressure hoses, balloons, balls, cushions, washers and gaskets. Water resistance has led to use in rainwear, diving gear, tubing and as linings for storage tanks and for road and rail tankers. Due to high electrical resistance, soft rubber is used as insulation and for protective gloves, shoes, and blankets; hard rubber is applied in articles such as telephone housings and parts for radios, meters and other electrical instruments.

Concern about the dependability of future global supplies of natural rubber arise from various factors, notably plant disease, climate change and the volatile market price of rubber, which affects investment in rubber plantations (Swain, 2021). For instance, in 2020 and 2021, the COVID-19 pandemic led to a surge in demand for rubber gloves and clothing, leading to a spike in rubber prices of about 30%.

SUMMARY

Natural rubber is an indispensable commodity in modern economies and demand for it continues to rise. It is a product from the plant kingdom that has had and continues to have a huge impact on human civilization (Smith, 2006).

Quite simply, it is difficult to overestimate the enormous economic and technological importance of rubber.

Questions

1. What is rubber? Describe different forms of rubber and explain its physical properties in relation to its chemical structure. Explain how knowledge of the molecular structure of natural rubber has helped to inform the manufacture of different synthetic rubbers with different repeat units in the polymer.
2. What is meant by thermoplastic and thermosetting as it is applied to polymers? Explain clearly the different applications of these types of polymeric compounds.
3. Why is natural rubber so vulnerable to oxidation, especially in the urban atmosphere when ozone, generated by sunlight acting on vehicle exhaust, might be present even in low parts per million?
4. Describe in an essay the importance in phytochemistry of the isomer and natural building block, isoprene.
5. What is vulcanization? Explain how it improves the commercial properties of natural rubber.

REFERENCES

J. E. Bowers. 1990. *Natural Rubber-Producing Plants for the United States*, pp. 11, 13. National Agricultural Library.

V. Cornelius-Takahama. 2001. *Sir Henry Nicholas Ridley*. Singapore Infopedia. Archived from the Original on 4 May 2013.

D. Hosler, S. L. Burkett, and M. J. Tarkanian. 1999. *Prehistoric Polymers: Rubber Processing in Ancient Mesoamerica'*. Science 284 (5422): 1988–1991. DOI: 10.1126/science.284.5422.1988. PMID 10373117.

L. W. Leng and K. J. Keong. 2011. *Mad Ridley and the Rubber Boom*. Malaysia History. Archived from the Original on 27 July 2013

C. Slack. 2002. *'Noble Obsession: Charles Goodyear, Thomas Hancock, and the Race to Unlock the Greatest Industrial Secret of the Nineteenth Century*. Hyperion.

J. P. Smith. Jr. 2006. *Plants & Civilization: An Introduction to the Interrelationships of Plants and People. Section 8.4, Latex Plants*, pp. 137–141. Humboldt State University Botanical Studies Open Educational Resources and Data, Humboldt State University Digital Commons.

F. Swain. 2021. *The Wonder Material We All Need But Is Running Out: Climate Change, Capitalism and Disease Are Threatening to Strike a Mortal Blow to the World's Rubber Trees. Do We Need to Find Alternative Sources of Rubber Before It's Too Late?*. BBC Future.

SUGGESTED FURTHER READING

W. Dean. 1997. *Brazil and the Struggle for Rubber: A Study in Environmental History*. Cambridge University Press.

T. Koyama and A. Steinbüchel, Eds. 2011. 'Biosynthesis of Natural Rubber and Other Natural Polyisoprenoids', in *Polyisoprenoids. Biopolymers*, Vol. 2, pp. 73–81. Wiley-Blackwell.

THE GREEN CREDENTIALS OF BIOPLASTIC

Abstract: *Plastics are ubiquitous in everyday life. Bioplastics, made from renewable plant resources, appear to be an attractive proposition offering environmental solutions, but are they quite as 'green' as they appear?*

Organic chemistry

- *polymerization*
- *carboxylic acids, alcohols and esters*
- *sugars*
- *condensation reaction*
- *chirality*

Context

- *biodegradability*
- *environmental issues*
- *sustainable practice.*

POLYMERIZATION AND PLASTICS

Plastics are produced by bonding together small molecules called monomers in a chemical reaction known as polymerization (see the chapters on 'Saving the Pacific Yew Tree'; 'Wheat—Ancient and Modern' and 'Asian Staple: Rice').

Most plastics, such as polythene or polystyrene, are made from expensive crude oil, which is processed industrially to produce various monomers for different plastics. Pollutants are produced in the process, such as CO_2, which contributes to climate change. Chemists have therefore looked at different ways to make plastics and avoid these two drawbacks. Hence the approach that involves the use of *renewable* plant material instead of crude oil to make bioplastics.

BIOPLASTICS

Bioplastics are materials produced from renewable biomass sources, such as corn starch, straw and woodchips, which can provide polysaccharides such as starch and cellulose or sugars, which yield lactic acid. Bioplastics can be made with a lower carbon footprint than their fossil-based counterparts, for example when biomass is used as both a raw material and as an energy source.

Polylactic Acid

Lactic acid is present in mammalian milk. A molecule of lactic acid has two functional groups characteristic of an aliphatic alcohol (hydroxyl) and a carboxylic acid (carboxyl; Figure 8.14). The formal name is alpha hydroxy propionoic acid. The alpha carbon atom is also chiral so there are two stereoisomers (enantiomers).

Lactic acid fermentation is a metabolic process that converts glucose into an aqueous solution of lactic acid with a release of energy. It is an anaerobic fermentation (oxygen free), which occurs in bacteria and in the muscle cells of animals where a build-up in human beings produces a sensation of muscular stiffness after exercise.

When lactic acid is polymerized in a condensation reaction, the product is known as polylactic acid (PLA). Although the name is in widespread use, it is inaccurate. As you can see from Figure 8.15, PLA is not a polyacid at all but is a polyester actually.

PLA is used for disposable consumer items such as plastic cups (Figure 8.16) and cutlery since it can be economically produced from renewable resources.

PLA is the most widely used plastic material in 3D printing (Figure 8.17). Its properties make it an ideal material for this purpose, namely low melting point, low thermal expansion, high strength, good adhesion plus high heat resistance once annealed.

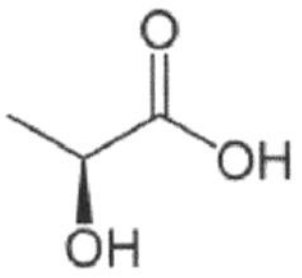

FIGURE 8.14   Lactic acid.

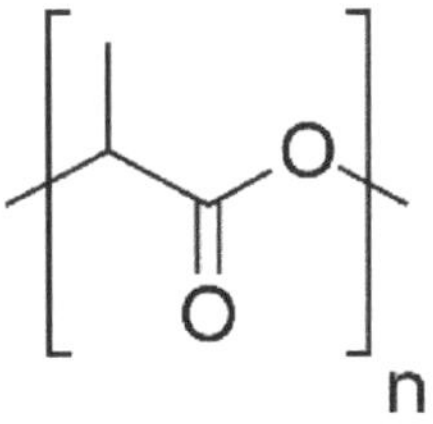

FIGURE 8.15   Formula of polylactic acid where *n* is a large number.

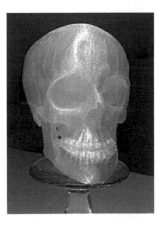

FIGURE 8.16   Disposable plastic cups made from PLA.

Source: cmglee Creative Commons Attribution-Share Alike 4.0 International license

FIGURE 8.17   A 3D-printed human skull in PLA.

Source: attribution Nevit Dilmen. Creative Commons Attribution-Share Alike 3.0 Unported License

There are medical applications too. PLA degrades into innocuous lactic acid, making it suitable for use in temporary medical implants that can break down inside the body within two years. This slow degradation gradually transfers the load to the body tissues as the area of repair heals.

According to American and European environmental standards, PLA is compostable but non-biodegradable. It does not biodegrade outside of the artificial composting conditions in waste disposal facilities.

Cellulose-Based Plastics

Cellulose bioplastics are mainly polymeric cellulose esters such as cellulose acetate (Figure 8.18).

As it is inflammable and flexible, cellulose acetate is used as the basis of film in photography. Cellulose acetate is also blended with yarns such as silk, cotton, wool and nylon, where it provides a fabric with excellent resistance to wrinkling, good dimensional strength and ability to be dyed in a pattern—all at a very competitive price. It can also be washed or dry cleaned and does not shrink. As it is made from made from wood pulp, a renewable resource, cellulose acetate can be composted or incinerated.

Lignin-Based Polymer Composites

Polymer composites based on lignin are bio-renewable and biodegradable. Lignin is a complex organic polymer that acts as a key structural material in the support tissues of most plants—particularly in cell walls (especially in wood and bark)—which provide rigidity and do not rot easily. Plants produce lignin from phenol precursors, which are then cross-linked in the polymer (Figure 8.19).

Lignin is a byproduct of polysaccharide extraction from plant material, especially wood, during the production of paper and ethanol—an essential feedstock of the chemical industry. As lignin is produced in great quantity, it is readily available for use as an emerging, environmentally friendly polymer. Lignin is also valuable since it has low weight and is neutral to CO_2 release during the biodegradation. Due to its distinctive chemical structure, lignin offers the potential for physical reinforcement in many composite polymeric materials.

Carbon polymers, such as graphene, are extensively applied in numerous fields: in energy storage and conversion and in various catalytic applications. Generally, carbon materials are derived from petroleum, which is non-renewable. Lignin is an ideal raw material as a carbon precursor due to low cost, high carbon content and its value in environmental protection (Dong, 2020).

RECYCLING TERMINOLOGY—LET'S BE CLEAR

A distinction between non-fossil-based plastic (*bioplastic*) and fossil-based plastic has limited relevance since materials such as petroleum are themselves merely fossilized biomass.

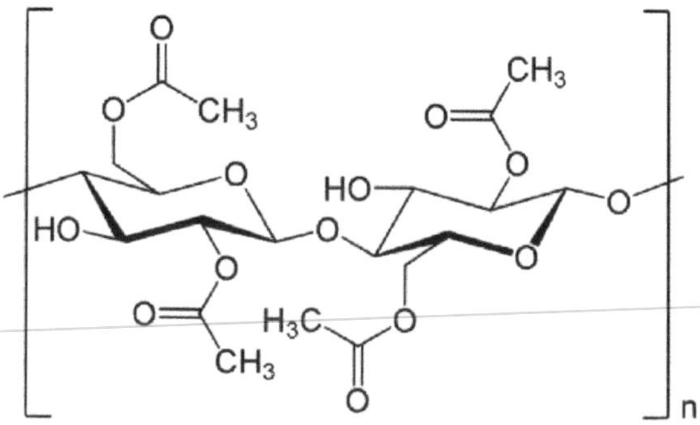

FIGURE 8.18   The repeat unit of cellulose acetate showing two acetyl groups on each module of glucose.

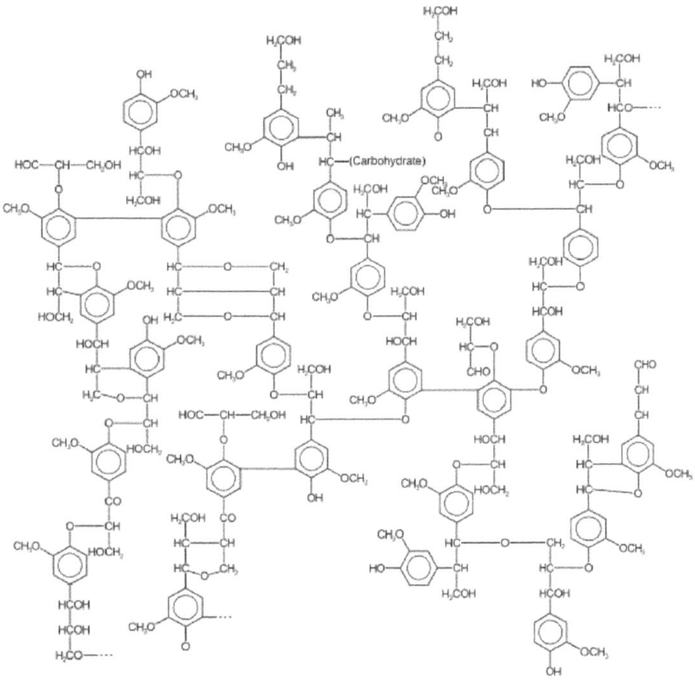

FIGURE 8.19 The cross-linked structure of lignin.

Source: Karol Glab. Creative Commons Attribution-Share Alike 3.0 Unported license

The term *bioplastic* itself may be misleading because it may be inferred that any polymer derived from the biomass is environmentally 'friendly'. The real question is whether a plastic is degradable or durable, that is, non-degradable. Not all bioplastic is biodegradable while some plastics produced from petroleum are biodegradable. The answer depends on the molecular structure of the plastic.

Biodegradable plastics are plastics that can be decomposed by the action of living organisms, usually microbes, into water, carbon dioxide and biomass. The conditions for this to happen may be artificially maintained in a waste disposal factory.

Compostable means plastic capable of undergoing biological decomposition in a compost site such that the material breaks down into carbon dioxide, water, inorganic compounds and biomass at a rate similar to that of natural materials.

All materials are inherently biodegradable, whether it takes a few weeks or a million years to break down into organic matter and mineralize. Credible companies convey the specific biodegradable conditions of their products, highlighting that their products are in fact biodegradable under national or international standards—anything less in consumer labeling may be considered as *'greenwashing'*.

In 2021, the European Commission conducted an evidence review on biodegradable plastics and concluded that:

> labeling plastic items as biodegradable, without explaining what conditions are needed for them to biodegrade, causes confusion among consumers and other users. It could lead to contamination of waste streams and increased pollution or littering. Clear and accurate labeling is needed so that consumers can be confident of what to expect from plastic items, and how to properly use and dispose of them.

Scientific advisors to the European Commission recommended 'coherent testing and certification standards for biodegradation of plastic in the open environment' (European Science Policy, 2021).

BIOPLASTICS AND THE ENVIRONMENT

Are bioplastics good for the environment? Well, let's take the single case of PLA as an illustration.

Hidden Production Costs

Some chemists point out hidden environmental costs: toxic pesticides sprayed on crops; carbon dioxide emissions from farm machinery and processing facilities and also energy use based on fossil fuels, but there are other concerns.

Effect on Food Supply

Replacing conventional plastic with polylactic acid derived from crop plants is not conceivable as the huge scale required would restrict food supply at a time when global warming is affecting farm productivity.

Methane Release

There is concern that methane, a potent greenhouse gas, will be released when biodegradable material, including truly biodegradable plastics, decomposes in landfill within an anaerobic environment.

$$2H_2O + C_6H_8O_4 \text{ (PLA repeat unit)} = 3CO_2 + 3CH_4$$

Of course, escaping methane could captured and used as clean, inexpensive fuel, but in so doing it would be converted into carbon dioxide.

$$CH_4 + 2O_2 = CO_2 + 2H_2O$$

Five Possible Scenarios for Disposal of PLA

Recycling can be by either chemical or mechanical means. Polylactic acid can be chemically recycled to monomer by heat or hydrolysis. When purified, the monomer can be used for the manufacturing of virgin PLA with no loss of original properties. End-of-life PLA can be chemically recycled to methyl lactate (Thompson, 2009). Generally, however, recycling is not straightforward. Products made from recycled plastic will only be structurally sound if one single type of plastic is involved. Recycling a mixture of different plastics is not feasible because each plastic has a different melting point and physical properties.

Composting: PLA is biodegradable under industrial composting conditions, starting with chemical hydrolysis process, followed by microbial digestion, to ultimately degrade the PLA.

Incineration: PLA can be incinerated because it contains only carbon, oxygen and hydrogen atoms. PLA can be combusted with no remaining residue indicating that incineration is an environmentally friendly disposal of waste PLA (Sultan, 2012).

Landfill: the least preferable option is landfill because PLA degrades very slowly in ambient temperatures and also may release methane in anaerobic conditions.

Biodegradation in the ocean: the oceans are not an optimal environment for biodegradation as the process is favored by warm conditions where microorganisms and oxygen are abundant. The deposition of plastic should be avoided. Indeed, microplastics, which are tiny remaining remnants of plastic that have not undergone full biodegradation but have merely been broken up by wave action and ultraviolet radiation in sunlight, are known to cause serious harm to marine life and may well be accumulating in the bodies of organisms higher in the food chain and, probably, in human beings too.

SUMMARY

While PLA accounts for about 50% of the bioplastic produced worldwide, bioplastics in total represent only about 1% of the plastic produced annually (European Bioplastics, 2023). New bioplastics of the future will have to possess properties that make them fully competitive with conventional plastics but will have far lower environmental impact than their predecessors. Ideally, bioplastics would

- be made from quickly growing, renewable plant resources
- be recyclable into other consumer products
- be biodegradable at least under industrial conditions in reclamation facilities
- be compostable, should any escape into the natural environment
- have little or zero carbon footprint using green sources of energy in production.

There would therefore be little environmental impact either from solid waste or greenhouse gases, CO_2 or CH_4. Of course, such a plastic would not be very durable and would be of lower economic value. There is clearly much work to do to ensure a balance between commercial viability and minimal environmental impact!

REFERENCES

H. L. Dong. 2020. *Preparation of Graphene-Like Porous Carbons with Enhanced Thermal Conductivities from Lignin Nano-Particles by Combining Hydrothermal Carbonization and Pyrolysis*. Front Energy Res 8: 148. DOI: 10.3389/fenrg.2020.00148.

European Bioplastics. 2023. www.european-bioplastics.org/market

Science Advice for Policy by European Academies (SAPEA). 2021. *Biodegradability of Plastics in the Open Environment*. Science Advice for Policy by European Academies. DOI: 10.26356/biodegradabiltyplastics

B. Sultan. 2012. *Global Warming Threatens Agricultural Productivity in Africa and South Asia*. Environ Res Lett 7 (4): 041001. DOI: 10.1088/1748-9326/7/4/041001.

R. C. Thompson. 2009. *Plastics, the Environment and Human Health: Current Consensus and Future Trends*. Phil Trans R Soc B 364 (1526): 2153–2166. DOI: 10.1098/rstb.2009.0053

SUGGESTED FURTHER READING

A. Barrett. 2018. *The History and Most Important Innovations of Bioplastic*. Bioplastics News.

A. Rudin and P. Choi. 2013. *Biopolymers*, pp. 521–535. The Elements of Polymer Science & Engineering. DOI: 10.1016/b978-0-12-382178-2.00013-4.

Questions

1. What are monomers, copolymers and repeat units?
2. What is meant by thermoplastic and thermosetting when applied to polymers? Give examples of the different commercial products for which the two types of polymer are used.
3. Explain the differences between the terms biodegradable and compostable as they apply to bioplastics.
4. Explain why some bioplastics may not necessarily be environmentally friendly products.
5. Describe and evaluate each of the various options available for dealing with waste PLA bioplastic.
6. About half of the bioplastic in commercial use is polylactic acid (PLA). Explain fully why PLA is such a favored bioplastic.
7. The alpha carbon atom in lactic acid is chiral. Explain what this means when lactic acid polymerizes to PLA.

Part IX

Plants and the Natural Environment

INTRODUCTION

Plants maintain the natural environment in many ways: releasing oxygen into the atmosphere, absorbing carbon dioxide, providing nutrients for animal life and helping to regulate natural cycles such as the water cycle and the nitrogen cycle. All are vital and fundamental to the natural processes sustaining life on Earth.

Since the 1850s when the anthropocene** began with the dawn of the industrial revolution, the essential and vital contribution of plants to sustainable practice and to measures to combat global warming assume ever greater significance and urgency with every year that passes.

A summary of the chemistry in Part IX follows

Chapter	Organic Chemistry	Context
Seaweed	Polymerization	Science of processed food
	Copolymers	Seaweed and soil fertilizers
	Carbon sequestration	Climate mitigation
	Polysaccharides	Alginic acid and alginates.
Garlic	Organo-sulfur compounds	Fossil fuels and air pollution
	Amino acids containing sulfur	Acid rain
	Scrubbing gaseous effluent	Scrubbing for Sulfur Dioxide
	Recycling sulfur	gases at industrial plants
Legumes	Nitrogen fixation	Soil fertility
	Nitrogen cycle	Haber process
	Inorganic synthetic fertilizer	Climate change

**The Anthropocene is the current geological age during which human activity has been a dominant influence upon climate and the environment.

DOI: 10.1201/9781032664927-9

SEAWEED AND 'BLUE' CARBON

Abstract: Seaweed, also known as macroalgae, is a general name for thousands of species of macroscopic, multi-cellular algae ubiquitous in the oceans of the world. Seaweeds take up CO_2 and release O_2 during photosynthesis. Seaweeds have long been exploited by man to enrich the soil for farming and as a source of food. Seaweed sequesters carbon effectively—a vitally important process known as 'Blue' carbon storage.

Organic chemistry

- *polymerization*
- *copolymers*
- *science of processed food*
- *seaweed and soil fertilizers*
- *carbon sequestration and climate mitigation*

Context

- *polysaccharides*
- *alginic acid and alginates.*

SEAWEED AND ALGAE

Seaweed is a very broad term. It refers to both flowering plants submerged in the ocean, like sea grasses but especially to large, marine, multi-cellular algae, which may be green, red or brown. Several thousand species of macroalgae are known. A few examples are shown in Figure 9.1.

Three environmental factors are necessary for seaweed to thrive: seawater (or at least brackish water), enough light for photosynthesis and a secure attachment point. While seaweed most commonly inhabits the littoral zone on rocky shores rather than on sand or shingle, there are some genera (*Sargassum* and *Gracilaria*) that float freely in open water. Some species of red algae live at great depths, where light levels are poorer, while other microalgae have adapted to live in shallow tidal rock pools in which rapidly changing temperature and salinity must be withstood (Lewis, J.R. 1964).

IMPORTANT NATURAL PRODUCTS DERIVED FROM ALGAE

Alginic acid is an edible polysaccharide with carboxyl functional groups along the molecular chain (see also the section on carbohydrates and polysaccharides in 'Asian Staple: Rice' in Chapter 3). A good source of alginic acid is brown algae, the largest species of which is giant kelp (*Macrocystis*

FIGURE 9.1   (a) Green seaweed (*Ascophyllum nodosum*).

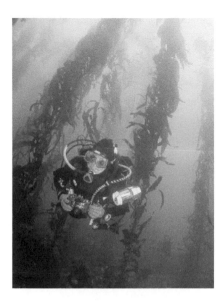

FIGURE 9.1   (b) Part of a forest of brown kelp.

Source: Aquaimages at English wikipedia. Creative Commons Attribution-Share Alike 2.5 Generic license

FIGURE 9.1   (c) Rhodophyta (red algae).

Source: Denisseacevedo073. Creative Commons Attribution-Share Alike 4.0 International license

pyrifera). Throughout the world, brown seaweeds are harvested to be processed and converted into *sodium alginate*. Brown seaweeds range in size from the giant kelp, which can be 20–40 m long, to smaller species 30–60 cm long. Most of the brown seaweed used for alginates is gathered from the wild, although, in China, some is cultivated for food with any surplus material being converted to sodium alginate. Alginic acid is a linear copolymer. The letters *m* and *n* denote sequences of different blocks of the two monomers involved (Figure 9.2).

The acid is hydrophilic but will absorb some water to form a viscous gum. Sodium and calcium alginate are salts of alginic acid, which are white to yellowish-brown solids that are used as a thickening agent in processed food.

Agar, sometimes referred to as agar-agar, is a jelly-like substance consisting of polysaccharides obtained from the cell walls of some species of marine red algae. Indeed, the name agar comes from agar-agar, the Malay word for red algae (Figure 9.3).

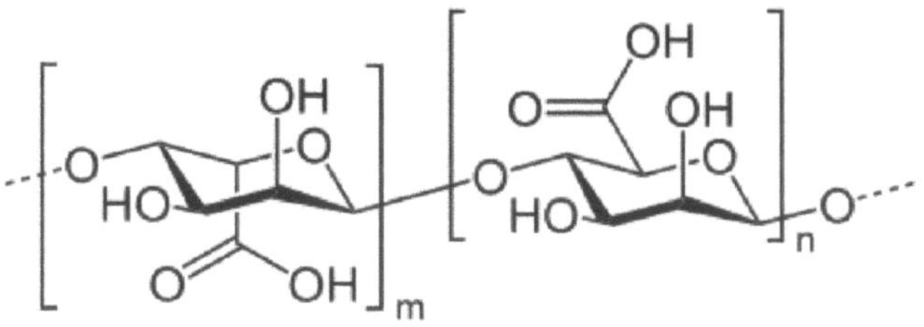

FIGURE 9.2 Alginic acid.

Source: image in the public domain

FIGURE 9.3 An example of red algae, Botryocladia occidentalis.

Source: Dagoberto E. Venera-Pontón, William E. Schmidt and Suzanne Fredericq. Creative Commons Attribution-Share Alike 4.0 International license

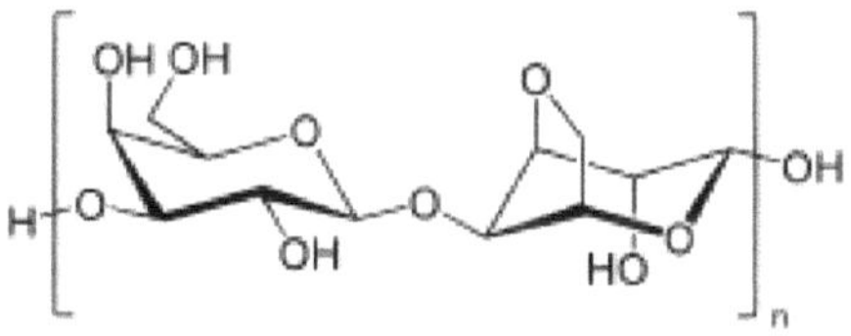

FIGURE 9.4 The structure of agarose.

Source: Yikrazuul. Public domain

Agar is a mixture of two polysaccharides: agarose being dominant in comprising 70% of the total. Agarose is a linear polymer (Figure 9.4), the repeat units being a disaccharide. The commercial product sold as agar is essentially pure agarose.

Agar will absorb water and solidifies between 32–42°C to form a gel that melts at 85°C.

VALUE TO MAN

Industry

Alginates are widely used in industrial products such as paper coatings, adhesives, dyes, gels and explosives. Seaweed alginate is an ingredient in toothpaste, cosmetics and paints.

Medicine

Dental and prosthetic impressions are taken with sodium alginate when artificial replacements are planned.

Food

Certain types of algae can be eaten. *Edible seaweed* is consumed across the world, particularly in East Asia. In Korea, Japan and China sheets of dried *Porphyra* are used in soups, sushi or rice balls. *Porphyra* is used in Wales to make laverbread. Alginate, agar and carrageenan are gelatinous seaweed products used as food additives where their gelling, water-retention and emulsifying properties are exploited.

In the *processed food industry*, agar is a permitted additive as a gelling agent, thickener, moisturizer, emulsifier and absorbent in many foodstuffs. Agar is used to make jellies, ice creams, puddings and custards. Sago is also produced from agar. Agar may be used instead of pectin in jams and marmalades due to its excellent property in gelling. For the same reason, agar is used to clarify stocks, sauces and other liquids. Agar does offer health benefits too as it is rich in dietary fiber.

Microbiology

In microbiology, a Petri dish containing agar provides a growth medium in which microorganisms, including bacteria and fungi, can be cultured and observed under the microscope (Figure 9.6). Many organisms cannot digest agar and so microbial growth does not affect the gel.

FIGURE 9.5 A treat in ice cream.

Source: Nicolas Ettlin. Creative Commons Attribution-Share Alike 4.0 International license

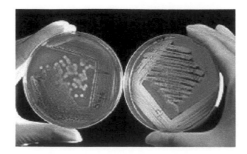

FIGURE 9.6   Petri dishes containing agar gel for bacterial culture.

Source: National Cancer Institute, USA released into the public domain

Since many scientific applications of agar require incubation of microbial cultures at temperatures close to human body temperature (37°C), agar is a more practical medium than gelatin derived from animal tissue, which melts at that temperature.

FARMING

Seaweed farming in its simplest form involves harvesting from natural beds, while at the other extreme the life cycle of the crop may be fully managed to improve yield. Seaweed can either be used to fertilize the soil or as a cash crop from aquaculture (Figure 9.7). Currently, the largest seaweed production occurs in China, Indonesia and the Philippines.

MITIGATION OF CLIMATE CHANGE

The inexorable rise in the concentration of CO_2 in the Earth's atmosphere is discussed in 'A Plant from the East Indies, Camphor' in Part VI. Seaweed cultivation, especially in the open ocean in aquaculture, could also make a significant contribution to mitigating climate change (Duarte et al., 2017; Temple, 2021). Seaweeds take up CO_2 and release O_2 during photosynthesis. Giant kelp (Macrocystis pyrifera) captures carbon faster than any other species of seaweed growing as rapidly as 0.5 m a day and reaching 60 m in length. According to one study, covering 9% of the world's oceans with kelp forests could produce 'sufficient methane to replace all of today's needs in fossil fuel energy, while removing billion tons of CO_2 per year from the atmosphere, restoring pre-industrial levels' (N'Yeurt et al., 2012; Buck, 2019). The IPCC Special Report on the Ocean and Cryosphere in a Changing Climate recommends 'further research attention' as a mitigation tactic (Bindoff, et al., 2019). World Wildlife Fund, Oceans 2050 and The Nature Conservancy in the UK publicly support expanded seaweed cultivation (Jones, 2023).

'BLUE' CARBON STORAGE

Seaweed detritus swept out to sea and then sinking to the deep ocean floor is an effective mechanism for sequestering carbon (Lavery p, 2018; Moore et al., 2023). It is a vitally important process that has become known as 'blue' carbon storage and is highly significant given that the oceans cover the majority of the surface of the Earth (Figure 9.8).

'Blue' carbon capture involves the use of marine ecosystems for carbon storage and removal. Seaweed aquaculture shows potential to act as a CO_2 sink through transformation of inorganic carbon into biomass during photosynthesis and the subsequent removal of that biomass for burial (Ortega, et al., 2019).

The use of seaweed as a fertilizer on land could also become an important contributor in climate mitigation strategy through avoidance of the use of artificial fertilizer with its attendant carbon

FIGURE 9.7   Aquaculture in the Philippines.

Source: Derek Keats. Creative Commons Attribution 2.0 Generic license

FIGURE 9.8   A polar view of the southern hemisphere of Earth reveals a vast expanse of ocean.

Source: image is in the public domain because it is a screenshot from NASA's globe software World Wind using a public domain layer, such as Blue Marble, MODIS, Landsat, SRTM, USGS or GLOBE

footprint (Raghunandan, 2019). The approach would also avoid run-off into rivers and oceans from fields dressed with artificial fertilizer. Seaweeds are also used as animal feeds. They have long been grazed by sheep, horses and cattle in Northern Europe. Adding seaweed to livestock feed substantially reduces methane emissions from cattle—methane being a potent 'greenhouse' gas (Irish Times, 2018).

Soil Chemistry and Seaweed Fertilizer

Organic Farming

Rising interest in organic farming practice is focusing attention upon introduction of seaweed-derived fertilizers rich in phyto-hormones and nutrients, polysaccharides, proteins and fatty acids, which improve soil and moisture retention contributing to higher crop yield (Patel HK, et al, 2019). Degradation of alginates also supplements the soil with organic matter, including fiber.

Brown seaweeds, such as *Sargassum*, are especially rich in alginates and alginic acid that improve soil quality by fostering populations of nitrogen-fixing bacteria, which also improve soil quality through their waste products (Zodape, 2001).

Seaweed fertilizer can bring about positive change in the physical condition of soil. Clay soils that lack organic matter benefit directly from the alginates found in seaweed. Also alginates chemically bond with calcium and magnesium atoms present in clay, stimulating its disaggregation, thereby improving the texture, aeration and porosity of the soil.

Remediation of Polluted Soils

Furthermore, seaweed can be used to remediate damaged soil. Functional groups (such as ester, hydroxyl, carbonyl, amino and phosphate groups) along the polymer chain of alginate drive the adsorption of heavy metal ions which are harmful pollutants (Kumar et al., 2006). Seaweeds such as *Gracilaria corticata varcartecala* and *Grateloupia lithophila* effectively remove a wide variety of heavy metals, including chromium (III) and (IV), mercury (II), lead (II) and cadmium (II) from their environment. Although significant potential exists for seaweed to remedy polluted soils, more research is needed to understand fully the mechanisms for these processes in the context of agriculture. It is possible that heavy metals accumulated by seaweed fertilizer could transfer to crops in some cases with significant implications for public health (Greger et al., 2007).

Summary

Perhaps it is easy to underestimate seaweeds as they are mostly submerged beneath the ocean waves, but they are vitally important plants. Quite apart from their value as foodstuffs and as soil fertilizer, seaweeds have the potential—if managed properly—to make a significant, global contribution to measures that secure climate mitigation. Macro and microalgae grow quickly, sequester CO_2 and have the capability to provide *renewable* feedstock for both the chemical industry and agriculture. These are considerations that are bound to attract significant attention in the years to come in a world where the norm will involve high regulation of CO_2 emissions, carbon accounting and related taxation.

Questions

1. What is 'blue' carbon? Explain the benefits and limitations of 'blue' carbon management.
2. What properties of alginates make them so attractive and useful to the food processing industry?
3. Explain at the molecular level why alginates can absorb water to form a gel.
4. Give a full account for the value of algal fertilizers in soil improvement for farming.
5. What are copolymers? Provide examples.
6. There is no single solution to deal with climate change but rather a suite of measures is required and then the measures need to be managed consistently. Describe complementary approaches to dealing with the principal greenhouse gas, carbon dioxide, which is building up in the Earth's atmosphere.

REFERENCES

N. L. Bindoff, W. W. L. Cheung, J. G. Kairo, J. Arístegui, et al. 2019. *Chapter 5: Changing Ocean, Marine Ecosystems and Dependent Communities*, pp. 447–587. IPCC Special Report on the Ocean and Cryosphere in a Changing Climate.

H. J. Buck. 2019. The *Desperate Race to Cool the Ocean Before It's Too Late*. MIT Technology Review.

C. M. Duarte et al. 2017. *Can Seaweed Farming Play a Role in Climate Change Mitigation and Adaption?* Front Mar Sci 4: 100. DOI: 103389/fmars.2017.00100.

M. Greger et al. 2007. *Heavy Metal Transfer from Composted Macroalgae to Crops*. Eur J Agron 26 (3): 257–265. DOI: 10.1016/j.eja.2006.10.003.

Irish Times. 2018. Seaweed Shown to Reduce 99% of Methane from Cattle. irishtimes.com. Retrieved 9 April 2018.

N. Jones. 2023. *Banking on the Seaweed Rush*. Hakai Magazine.

V. V. Kumar et al. 2006. *Biosorption of Metals from Contaminated Water using Seaweed*. Curr Sci 90 (9): 1263–1267. DOI: 10.13140/2.1.2176.4809.

O. W. Moore et al. 2023. *The Maillard Reaction Helps to Store Carbon on the Sea Floor*. Nature. DOI: 10.1038/s41586-023-06325-9.

A. R. N'Yeurt et al. 2012. *Negative Carbon via Ocean Afforestation*. Proc Saf Environ Protect. Special Issue: Negative Emissions Technology 90 (6): 467–474. DOI: 10.1016/j.psep.2012.10.008.

A. Ortega, N. R. Geraldini, A. A. Kamau, and C. M. Duarte. 2019. *Important Contribution of Macroalgae to Oceanic Carbon Sequestration*. Nature Geoscience 12: 748–754 (2019).

B. L. Raghunandan. 2019. *Perspectives of Seaweed as Organic Fertilizer in Agriculture*, pp. 267–289. Soil Fertility Management for Sustainable Development.

J. Temple. 2021. *Companies Hoping to Grow Carbon-Sucking Kelp May Be Rushing Ahead of the Science*. MIT Technology Review. CS1 maint: url-status (link)

S. T. Zodape. 2001. *Seaweeds as a Biofertilizer*. J Sci Ind Res 60 (5): 378–382.

SUGGESTIONS FOR FURTHER READING

V. J. Chapman and D. J. Chapman. 1980. *Seaweeds and Their Uses*, 3rd Edition, p. 148. Springer.

A. Davidson. 2006. *The Oxford Companion to Food*. Oxford University Press.

J. R. Lewis 1964. *The Ecology of Rocky Shores*. The English Universities Press Ltd.

S. Liu and L. Usinger. 2008. *Agar and Its Use in Chemistry and Science*. Science Buddies.

C. S. Lobban and M. J. Wynne. 1981. *The Biology of Seaweeds*, pp. 734–735. University of California Press.

National Oceanic and Atmospheric Administration, USA. 2021. *How much Oxygen comes from the Ocean?*

GARLIC AND PUNGENT SMELLS

Abstract: *Garlic, known as Allium sativum, is a species in the onion genus, Allium. Garlic has been in use for both culinary and medicinal purposes for over 7,000 years.*

Organic Chemistry

- *organo-sulfur compounds*
- *their influence upon environmental pollution*
- *preventing the escape of unwanted sulfur dioxide into the atmosphere from industrial processes*
- *acids and bases*

Context

- *sulfur compounds in natural products.*

GARLIC

Garlic, known as *Allium sativum* (Figure 9.9), is a species in the onion genus. Garlic is frequently used in cooking and in food preparation but its use is associated with the socially undesirable reputation of 'garlic breath'. The compound that leads to garlic odor is not present in fresh garlic but is formed only when garlic is crushed or minced. This action causes enzymes to break down a natural compound named alliin to form allicin, which contributes to the familiar odor of crushed garlic.

ORGANO-SULFUR COMPOUNDS

Organo-sulfur compounds are common in nature and are often associated with foul smells when chemical breakdown leads to the release of hydrogen sulfide and/or ammonium sulfide.

Indeed, several organo-sulfur compounds are present in garlic, examples being diallyldisulfide and ajoene (Figure 9.10). Other sulfur-containing organic compounds are diallyl disulfide, allyl methyl sulfide, allyl mercaptan, and allyl methyl disulfide. Of these, allyl methyl sulfide is the compound that takes longest for the body to break down. It is absorbed in the gastrointestinal tract, passes into the bloodstream and moves on to other organs in the body; for excretion through the skin via sweating, via the kidneys in passing urine and is also exhaled from the lungs. Also present in these organo-sulfur compounds is a functional group known as the allyl group. For more on the allyl function, see lycopene in the chapter on 'Saffron and carotenoids—yellow and orange dyes' in Part VII.

FIGURE 9.9   Garlic bulbs, *Allium sativum.*

Sulfur-containing compounds are responsible for the antibacterial properties of garlic. These compounds penetrate the membranes of a bacterium cell where they cause changes in the structures of enzymes and proteins, which contain the functional group known as a thiol (-SH), thereby injuring the cell.

AMINO ACIDS THAT CONTAIN SULFUR

In the chapter on 'Wheat—Ancient and Modern' in Part III, we introduced essential amino acids. Human beings need twenty essential amino acids for good metabolism and the formation of proteins. Two of these essential amino acids, namely methionine and cysteine, contain a sulfur atom (Figure 9.11). Cysteine is found in garlic and onions.

A molecule of cystine (Figure 9.12) can be produced when two molecules of cysteine interact through the formation of an S-S bond. This type of covalent bond is important in cross-linking

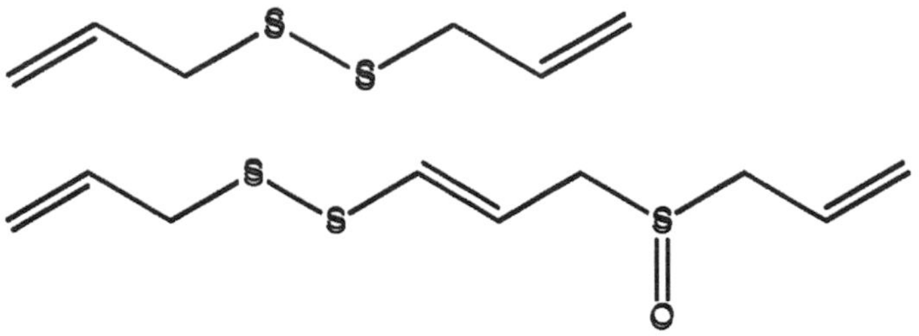

Diallyldisulfide (DADS)

Ajoene

FIGURE 9.10 Two important compounds containing sulfur found in garlic, diallyldisulfide (DADS) and ajoene.

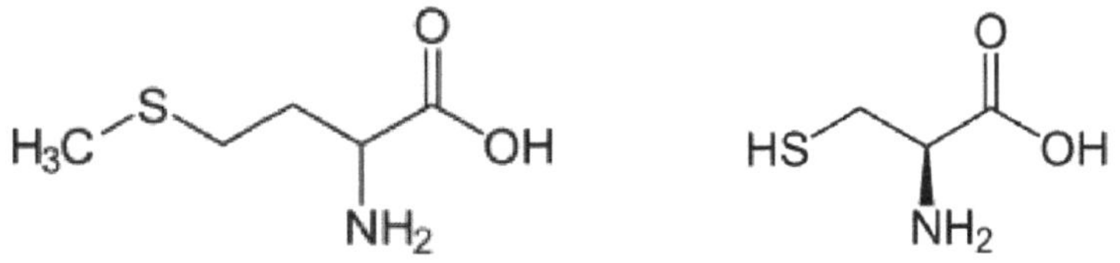

Methionine

Cysteine

FIGURE 9.11 Two important amino acids that contain sulfur.

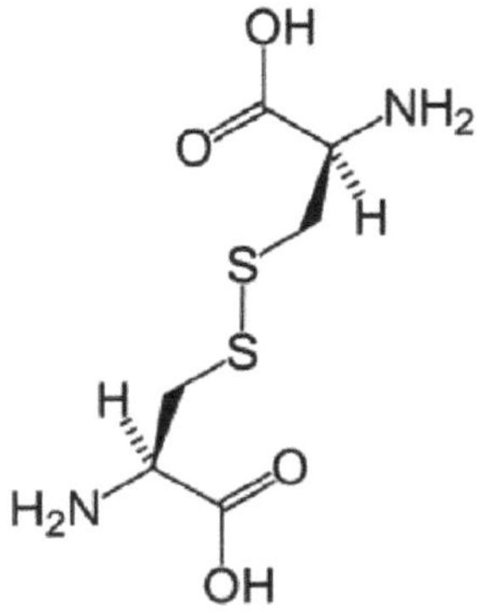

FIGURE 9.12 Cystine.

protein molecules. The other two ways of cross-linking proteins arise from hydrogen bonding and ionic bonding.

S–S bonds are also prominent in another way in natural products chemistry in that they can be readily involved in redox reactions (see Glossary). One such example occurs when human hair is restyled, which is explained in chemical terms in the following manner.

Human hair and skin contain approximately 10% of cystine by mass. At the barber's parlor, singed hair characteristically releases some hydrogen sulfide. Also, hair treatments involve the breaking of the sulfur–sulfur bond allowing the hair to be re-formed in a different style.

Hair is made mostly of a protein called keratin, which is also present in nails. In hair, keratin molecules are arranged in parallel bundles, which are bound together by cross-linking disulfide bonds. Cysteine, present in one keratin molecule, forms a disulfide bond with the cysteine of a neighboring keratin molecule. In this manner, the greater the number of disulfide linkages present in a strand of hair, the straighter the hair becomes.

Ammonium thioglycolate ($HSCH_2CO_2NH_4$) is a compound that can break disulfide bonds as it contains a thiol functional group. As the disulfide bonds are broken by chemical reduction to thiol groups, the strands of keratin come apart. When hair is restyled with curls, it is called a perm (an abbreviation of permanent waving). Straightening hair is called re-bonding. In both cases, the steps are very similar. Once the hair has been washed to clean it thoroughly, ammonium thioglycolate solution is applied for a short while. If a perm is desired, hair is tied around curlers. If the hair is to be re-bonded, it is pressed firmly among flat irons until it becomes straight.

When the hair has been shaped, the strands of keratin need to be re-connected so that the style is retained. An oxidizing lotion is applied, containing hydrogen peroxide, which reconstitutes the disulfide bonds. Hence, the process involves a redox reaction—also described in the chapter on 'Cocoa: Food of the Gods' in Part IV.

Fossil Fuels and Acid Rain

Fossil fuels (coal, petroleum and natural gas) are the decayed remains of ancient organisms and contain, as a consequence, a small proportion of organo-sulfur compounds. The combustion of these compounds by man leads to an accumulation of sulfur dioxide in the atmosphere, which is a major component of air pollution as the gas combines with water to form droplets of sulfuric acid. These droplets fall as acid rain, which harms and kills plants and animals, especially those that live in aquatic environments. Acid rain also damages buildings faced in marble or limestone by direct chemical action:

$$H_2SO_4 + CaCO_3 = CaSO_4 + CO_2 + H_2O$$

Sulfur and oxygen are both in Group VI of Mendeleev's classification of the elements, the Periodic Table. As a consequence of the chemical similarity between sulfur and oxygen, organic compounds containing carbon-sulfur and carbon–oxygen bonds have some similar properties—examples being the alkyl thiols, R-SH, and the alkyl alcohols, R-OH.

Scrubbing Gaseous Effluent Gases at Industrial Plants

At power stations where fossil fuels are still being used to generate electricity, it is clearly necessary to operate scrubbers in exhaust chimneys to remove sulfur dioxide (SO_2) from effluent gases.

As (SO_2) is an acid gas, slurries used to remove SO_2 from flue gases are usually metal carbonates or alkaline metal hydroxides.

The reaction taking place when calcium carbonate or limestone ($CaCO_3$) is used produces calcium sulfite ($CaSO_3$)

$$CaCO_{3(s)} + SO_{2(g)} \rightarrow CaSO_{3(s)} + CO_{2(g)}$$

FIGURE 9.13   A scrubbing tower at an industrial plant.

whereas scrubbing with hydrated lime or calcium hydroxide ($Ca(OH)_2$) produces calcium sulfite ($CaSO_3$).

$$Ca(OH)_{2(s)} + SO_{2(g)} \rightarrow CaSO_{3(s)} + H_2O_{(l)}$$

A solution of caustic soda or sodium hydroxide (NaOH) is also often used to scrub SO_2 producing sodium sulfite.

$$2\,NaOH_{(aq)} + SO_{2(g)} \rightarrow Na_2SO_{3(aq)} + H_2O_{(l)}$$

An Alternative Method of Reducing Sulfur Dioxide Emissions

An economic alternative to removing sulfur from flue gases after burning is to remove sulfur from the fuel either before or during combustion by adding lime. Calcium sulfate is formed, some of which is retained and purified thereby reducing the need to mine gypsum rock. Sulfur may be also recovered from ash residue and recycled for use elsewhere. The whole process takes place at atmospheric pressure and ambient temperature (Paquell BV. 2019).

Questions

1. Compare the molecular structures of diallyl disulfide and ajoene presented in the figure earlier in the chapter. Where do you think that the weakest points in the chains of each these molecules occur and why?

2. Give an account of other environmental effects of sulfur dioxide pollution due to increased pH levels in rain and lakes.

3. Describe the measures taken at power stations which use fossil fuels to prevent sulfur dioxide from entering the atmosphere.

4. Apart from power stations, what other sources of air pollution by sulfur dioxide are there and what preventive measures may be taken?

5. Sulfur and oxygen are both in Group VI of Mendeleev's classification of the elements, the Periodic Table. Explain why sulfur and oxygen are found in the same chemical group in the Periodic Table.

6. As a consequence of the chemical similarity between sulfur and oxygen, organic compounds containing carbon–sulfur and carbon–oxygen bonds have some similar properties. Compare and contrast an alkyl alcohol, R-OH, with an alkyl thiol, R-SH.

SUGGESTED FURTHER READING

H. P. Koch and L. D. Lawson, Eds. 1996. *Garlic. The Science and Therapeutic Application of Allium Sativum and Related Species*, 2nd Edition. Williams & Wilkins.

Paqell BV. 2019. *HIOPAQ Oil and Gas Process Description*. Paqell BV. Paqell is a joint venture between Shell Global Solutions and Paques.

LEARNING TO LOVE LEGUMES

Abstract: Legumes have been cultivated for millennia as sources of both human nutrition and fodder for animals. As a natural fertilizer, legumes also play a key role in crop rotation.

Legumes are notable in that most of them have a symbiotic relationship with nitrogen-fixing bacteria which are present in nodules on the roots.

Both as a food source and a natural soil fertilizer, legumes make a vital contribution of global significance toward offsetting the unsustainable dependence of modern agriculture upon inorganic fertilizer produced industrially using energy from non-renewable fossil fuels.

Organic chemistry

- *nitrogen fixing*
- *nitrogen cycle in nature*

Context

- *legumes as a natural soil fertilizer*
- *Haber process and unsustainable production of inorganic nitrogen fertilizer.*

LEGUMES

A legume is a plant in the large family *Fabaceae* (also known as *Leguminosae*), which extends to clover, soybean, alfalfa, peanuts, beans, peas and lupin. Dried seed is known as pulse (Figure 9.14). Legumes are grown agriculturally for human consumption, for livestock forage and silage and as soil-enhancing, green manure. As a natural fertilizer, legumes play a key role in crop rotation.

NITROGEN FIXING

Legumes are notable in that most of them have a symbiotic relationship with nitrogen-fixing bacteria. Legumes co-exist with bacteria, located in small nodules on their roots, to fix nitrogen from the air

FIGURE 9.14   Freshly picked legumes and dried pulses.

Source: image is in the public domain because it contains materials that originally came from the Agricultural Research Service, the research agency of the United States Department of Agriculture.

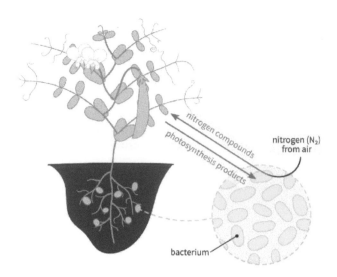

FIGURE 9.15   Illustration of nitrogen fixation.

Source: Nefronus. Creative Commons Attribution-Share Alike 4.0 International license

(Figure 9.15). Nitrogen is an essential nutrient required by all plants for growth and development and is usually assimilated from the soil in the form of nitrates of ammonia, sodium and potassium, which are all soluble in water.

Rhizobia bacteria chemically combine nitrogen gas (N_2) to ammonia (NH_3) in a process called nitrogen fixation. The plant then converts the nitrogen compounds arising from the ammonia in the root nodules into the amino acids, vitamins and flavones that are essential for growth. The plant supplies the energy required to cleave nitrogen molecules from sugar transferred from leaves where sugars are made through photosynthesis. As the sugar, sucrose, breaks down, esters of malic acid are formed, which are a direct source of carbon for the bacteria. Root cells in the leguminous plant release the organic acid to the bacteria in exchange. Hence, the legume and bacterium have a symbiotic relationship from which each organism benefits.

Malic acid is a dicarboxylic acid that contributes to the sour taste of fruit—especially apples. Malic acid has two stereoisomers, L- and D-enantiomers, though only the L-isomer exists in nature.

When the plant dies, organic compounds containing nitrogen fixed from the air are released into the soil providing an additional source of fertilizer for other plants. Traditional farming practice takes advantage of this. Over the years, fields are rotated through various types of main crop—one of which will be a leguminous plant such as clover that is ploughed in at the end of the growing season.

NITROGEN CYCLE

Nitrogen fixation from the atmosphere by legumes, shown on the left of the illustration in Figure 9.17, is an important part of the natural nitrogen cycle.

The water-soluble nitrate salts produced are eventually absorbed by green plants, assisting their development, from whence the whole cycle starts again.

INDUSTRIAL PRODUCTION OF NITROGEN FERTILIZER

As we have seen, nitrogen fixation from the air by leguminous plants is a valuable, natural process of soil enrichment.

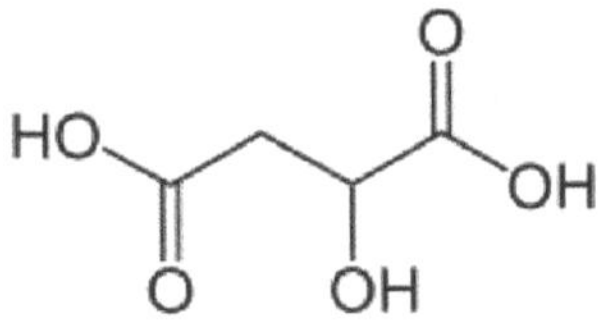

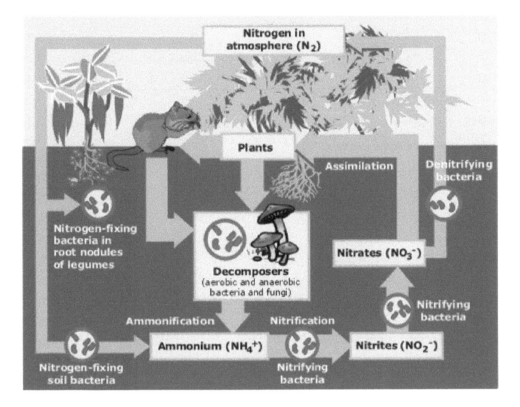

FIGURE 9.16   Malic acid.

FIGURE 9.17   The nitrogen cycle.

Source: Environmental Protection Agency. As works of the U.S. federal government, all EPA images are in the public domain.

Nonetheless, throughout the 19th century, a principal source of soil fertilizers were the inorganic compounds, sodium and potassium nitrate, which were obtained directly from huge deposits of guano (seabird droppings) present on the surface of the arid Atacama Desert in Chile. However, it became clear that these natural reserves could not satisfy future demand and so research into new sources of inorganic fertilizer based on ammonia became important. Once manufactured, ammonia can easily be converted into ammonium nitrate (NH_4NO_3), a water-soluble inorganic fertilizer containing nitrogen.

Ammonia has now become an extremely valuable commodity used in a huge range of commercial products from cleaning agents to plastics production to refrigerants and, in particular, to produce inorganic nitrate fertilizer. Ammonia is manufactured industrially on a vast scale with global production reaching around 240 million tons in 2020 (Figure 9.18).

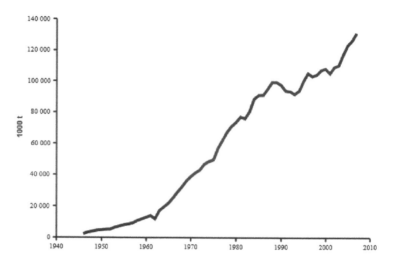

FIGURE 9.18   Industrial production of ammonia following the Second World War in million tons.

Source: Ammoniakproduktion.svg: Orci

FIGURE 9.19   Wheat harvest.

Source: Creative Commons Attribution-Share Alike 3.0 Unported license. Michael Gäbler/Wikimedia Commons/CC BY-SA 3.0

https://commons.wikimedia.org/w/index.php?curid=28285629

Most of the ammonia goes toward making fertilizer and, significantly, much of the world's food production is dependent upon it (Figure 9.19).

An obvious source of nitrogen (N_2) is Earth's atmosphere. However, N_2 is exceptionally stable. Production of ammonia directly from nitrogen represented a serious challenge. Fritz Haber studied the effects of pressure, temperature and catalysis on the process in the laboratory in 1909. Carl Bosch, an employee of the German chemical company Badische Anilin und Soda Fabrik (BASF), was given the task of scaling up Haber's reaction to industrial level, which resulted in BASF manufacturing ammonia in 1913. Haber and Bosch were later awarded Nobel prizes for their work in overcoming the chemical and engineering problems involved.

The *Haber Bosch process* involves the conversion of nitrogen and hydrogen into ammonia at high temperature (500°C) and pressure (200 atmospheres) in a reversible reaction.

$$N_2(g) + 3H_2(g) = 2NH_3(g)$$

Fritz Haber realized that the speed of the forward reaction depended largely upon the fact that the triple covalent bond in nitrogen is really hard to break. He experimented by externally applying pressure and temperature to the dynamic equilibrium and also discovered that iron could act as a catalyst to allow the reaction to occur much more quickly.

Although the Haber Bosch process represented huge technological advancement, its energy demands result in a massive carbon footprint (Department of Business, Energy and Industrial Strategy, 2020). Hydrogen used in the process comes from methane (CH_4) obtained from fossil fuels (natural gas, coal and oil) through processes that release CO_2. From hydrocarbon feedstock through to NH_3 synthesis, it is estimated that every NH_3 molecule generated incidentally releases one molecule of CO_2. Since modern agriculture depends in a big way upon nitrate fertilizers based on ammonia, a significant part of our food is effectively a product derived unsustainably from fossil fuels.

The Haber Bosch process is over 100 years old and is expensive in energy consumption, yet no industrial-scale alternative exists, and so it remains in use today. High emissions of the potent greenhouse gas, carbon dioxide, provide strong impetus for modern research effort into other processes that may use sources of green electricity.

ALTERNATIVE, SUSTAINABLE ROUTES FOR SYNTHESIZING NITROGEN FERTILIZER

Over the past two decades, alternatives to the existing Haber Bosch process have been investigated among them *direct* electrochemical synthesis in aqueous, molten salt or solid-state electrolytic cells (Marnellos, 1998 and Garagounis et al., 2019).

Another process definitely known to produce ammonia was first reported in the Japan during the 1990s and involves *lithium nitride*. Lithium is the smallest and most reactive atom of the Group 1 alkali metals. Lithium reacts exothermically with nitrogen gas at room temperature forming crystalline lithium nitride (Li_3N; Figure 9.20).

Lithium will even react readily with nitrogen in the air to give lithium nitride.

$$6Li + N_2 \rightarrow 2Li_3N$$

Lithium nitride is a stable reddish-pink crystalline solid that reacts violently with water to produce ammonia:

$$Li_3N + 3H_2O \rightarrow 3LiOH + NH_3$$

Scientists at Stanford University in California have recently devised a cyclic, electrochemical process to make ammonia involving lithium nitride and report an efficiency of 88.5% in the laboratory. Electricity from a renewable source is used to electrolyze lithium hydroxide and produce lithium metal.

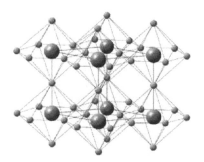

FIGURE 9.20 Polyhedral structure of crystalline lithium nitride with nitrogen atoms shown in orange.

Exposing this lithium to a stream of nitrogen gives lithium nitride. Lithium nitride is then hydrolyzed to produce ammonia and regenerated lithium hydroxide. The process works at ambient pressure, moderate temperature and importantly does not require the expensive production of hydrogen (McEnaney et al., 2017). Research continues into the properties of cheaper metal nitrides, such as magnesium nitride (Mg_3N_2), which can produce ammonia in a similar way.

However, means have not yet been found to scale-up significantly any of these reactions into an industrial process capable of replacing the Haber Bosch method so as to meet huge worldwide demand for ammonia and for the inorganic nitrogen fertilizer produced from it.

SUMMARY

One of the greatest challenges of the present day is de-carbonization of the global economy and finding an alternative to the Haber Bosch process would certainly help.

In the meantime, the reliable, extensive, twin contributions of leguminous plants as both a nutritious source of food and as an entirely natural, organic, nitrogen-based soil fertilizer should not be underestimated.

Questions

1. Explain the process of nitrogen fixation from the atmosphere.
2. Describe the key environmental issues that arise from the industrial production of inorganic nitrogen fertilizer.
3. Give an account of the contribution of legumes in offsetting agricultural demand for inorganic soil fertilizer.

REFERENCES

Department of Business, Energy & Industrial Strategy. 2020. *UK Energy Statistics*. Quarterly Bulletin.
I. Garagounis et al. 2019. *Electrochemical Synthesis of Ammonia: Recent Efforts and Future Outlook.* Membranes 9 (9): 112. DOI: 10.3390/membranes9090112.
G. Marnellos. 1998. *Ammonia Synthesis at Atmospheric Pressure*. Science 282: 98–100. DOI: 10.1126/science.282.5386.98.
J. M. McEnaney et al. 2017. *Ammonia Synthesis form Nitrogen and Water using Lithium Cycling Electrification at Atmospheric Pressure*. Energy Environ. Sci. DOI: 10.1039/c7ee01126a

Glossary

Aliphatic Compounds: the term aliphatic compound refers to any open-chain hydrocarbon such as an alkane, an alkene and an alkyne.

Alkaloids: alkaloids are a group of naturally occurring chemical compounds that contain nitrogen atoms within cyclic rings of carbon and hydrogen atoms. The nitrogen atoms are chemically basic. Alkaloids may also contain oxygen and sulfur. Alkaloids are found widely in nature being produced by a large variety of organisms: bacteria, fungi, plants and animals and are often *toxic* to other organisms. Although many alkaloids have a bitter taste they have been found to have pharmacological effects in human beings. As a consequence, they are used for medication or as recreational drugs: examples being the local anesthetic and stimulant cocaine, the stimulants caffeine and nicotine, the analgesic morphine, the anti-cancer compound vincristine and the anti-malarial drug quinine.

Antigen: an antigen is a molecule that is foreign or toxic to an organism (a bacterium or a virus) that induces an immune response. An antigen attracts and is bound to a specific antibody prepared by the organism to deal solely with that antigen.

Apoptosis: apoptosis is name given to a process of programmed cell death that can occur in multi-cellular organisms. Cell death follows a controlled path. In general, apoptosis confers advantages during an organism's lifecycle. It is certainly essential for human development; for example, the separation of fingers and toes in a developing human embryo occurs because cells between the digits undergo apoptosis. Apoptosis is also necessary at the start of menstruation.

Research into the effects of apoptosis has increased substantially in recent years because of links with a variety of diseases. Excessive apoptosis causes atrophy as in Alzheimer's disease and Parkinson's disease, whereas an insufficient degree of apoptosis can result in uncontrolled cell proliferation or cancer.

ATP: **a**denosine **trip**hosphate (ATP) is an organic substance found in all living cells that is a source of energy and acts rather like an enzyme. ATP transports chemical energy within a cell to allow metabolism to occur. It is important to note that metabolic processes that utilize ATP later restore it. ATP is therefore continuously recycled in organisms.

Building Block: a building block is a term used in organic chemistry that is used to describe a molecule of an organic chemical compound that has one or more active functional groups. As the term implies, building blocks may be assembled through reactions involving the functional groups to form much bigger or more complex molecules.

Chelate: a chelate is a compound containing a ligand bonded to a central metallic ion at one or more points. The International Union of Pure and Applied Chemistry (IUPAC) define chelates as compounds which involve the presence of two or more separate co-ordinate bonds between a polydentate (multiple bonded) ligand and a single central atom or ion. See also the definition of a ligand.

Chirality: a molecule or ion is called *chiral* if it cannot be superposed on its mirror image. This geometric property is called *chirality*. A chiral molecule or ion exists in two *stereoisomers* that are mirror images of each other, called *enantiomers*, often distinguished as either 'right-handed' or 'left-handed' due to their configuration. The two enantiomers have the same chemical properties, except when reacting with other chiral compounds, and they have the same physical properties, except that they often rotate polarized light in opposite directions. An homogeneous mixture of two enantiomers in equal parts is described as *racemic*. The racemic mixture usually differs chemically and physically from pure enantiomers.

Chloroplast: the chloroplast is the place within the cell of a green plant where photosynthesis occurs.

Chromophore: a chromophore is the part of a molecule responsible for its color. The color arises when a molecule *absorbs* certain *wavelengths* of visible light and transmits or reflects others. The chromophore is a region in the molecule where the energy difference between two different molecular orbitals falls within the range of the visible spectrum. Visible light that hits the chromophore can thus be absorbed by exciting an electron from its ground state into an excited state.

Conformer: conformational isomerism occurs through rotation about a single bond in a molecule. It differs from structural stereo-isomerism where inter-conversion would require the breaking and reforming of chemical bonds.

Conjugation: in chemistry, conjugation occurs in compounds with alternating single and multiple bonds, which allows p-orbitals to interconnect and delocalize electrons thereby increasing stability in the molecule. A conjugated compound may be linear, cyclic or mixed.

Critical Point: see the entry for supercritical fluids.

Dalton: a Dalton is a physical constant accepted within the international SI system that has been named after the famous British chemist, John Dalton, who first proposed its use. The Dalton (represented as u or Da) is a relative unit of mass being based upon one twelfth of the mass of a carbon nucleus. The actual mass of 1u or Da is $1.6605389217 \times 10^{-27}$ kg.

Diastereomers and Enantiomers: diastereomers (also known as diastereoisomers) arise when two or more stereo-isomers of a compound exist that have different configurations at one of the equivalent stereo-centers and which are not mirror images of each other. Stereo-isomers that are mirror images and may be superimposed on one another will rotate the plane of polarized light and are known as enantiomers.

Dimers: a dimer consists of two monomers joined by bonds that can be strong or weak, covalent or intermolecular. Dimers can have significant implications in polymer chemistry; inorganic chemistry and biochemistry.

DNA: **d**eoxyribo**n**ucleic **a**cid is a very long polymer made up from monomers called nucleotides. Nucleotides are in themselves complex molecules but along a DNA polymer there are only four kinds of nucleotide, each one containing a different base: adenosine (A), cytosine (C), guanine (G) or thymine (T). These four different nucleotides can occur in any sequence along the DNA chain. It is the order that contains the information representing the unique genetic code of an organism.

Electrophile: an electrophile, such as a proton or a nitroxide cation, is an electron acceptor that is attracted to electron-rich parts of molecules and often undergoes a chemical reaction there.

Elimination Reactions: in an elimination reaction, a small group of atoms or a small molecule break away from a larger molecule and are not replaced.

Enzyme: an enzyme is a biological catalyst that speeds up or facilitates chemical reactions in organisms.

Enzyme Inhibitor: an enzyme inhibitor is a molecule that binds to an enzyme and decreases its activity.

Functional Group: in organic chemistry, a functional group is a specific group of atoms or bonds within a molecule which is responsible for characteristic chemical reactions. However, the relative reactivity of a functional group may be modified by other functional groups nearby and by the structure of the molecule.

Hodgkin's disease: Hodgkin's disease appears as a tumor in the glands known as lymph nodes found in the neck, groin, armpits and chest. The lymphatic system is a network of vessels that runs throughout the body carrying the colorless fluid, lymph, which transports the white blood cells, so important in maintaining immunity from infection. Abnormal white cells are produced by the lymph nodes and are a characteristic of Hodgkin's disease.

Hormones: a hormone is a regulatory substance produced in an organism and transported in tissue fluids such as blood or sap to stimulate specific cells or tissues into action.

Hydrogen bonding: covalent bonds consisting of a hydrogen atom bonded to an atom of a strongly electronegative element (such as oxygen, nitrogen or a halogen) are highly polar. These polar bonds give rise to permanent dipoles, which result in electrostatic attraction between neighboring molecules—known as hydrogen bonding.

Hydrophilic: meaning water 'loving'.

Hydrophobic: meaning water 'hating'.

Le Chatelier's principle: if a system is at equilibrium and a change is made in any of the conditions, then the system responds to counteract the change as much as possible

Lipids: lipids are fatty acids that are insoluble in water yet soluble in many organic solvents. The class includes natural oils, waxes and steroids.

Ligand: compounds exist in which a central metal ion co-ordinates with electron-rich groups called ligands. The aquated ion of copper is an example, $(Cu(H2O)6)2+$, which gives copper sulfate solution its deep blue color—anhydrous copper sulfate being a white powder.

Metabolites: metabolites are intermediates in metabolic processes in nature and are usually small molecules.

Miscible: the term miscible applies when two or more liquids, such as water and ethanol, are able to dissolve into one another in any proportion without separating.

Mitochondria: in mitochondria within cells, nutrients are broken down to release energy. Enzymes in the mitochondria act in aqueous solution to oxidize molecules of food. The process is known as cellular respiration.

Mitosis: in mitosis, a cell divides to create two identical copies of the original cell.

Mole: one mole is defined as the molecular weight of a substance expressed in grams based on the standard of 12 grams of carbon 12.

Nucleophile: a nucleophile is an electron-rich species such as an anion that undergoes chemical interaction with a molecule by being attracted to atoms in dipolar bonds that have a positive charge.

Primary Metabolites: primary metabolites are needed for the growth, development and reproduction of an organism.

Phototropism: phototropism is the orientation of a plant or other organism in response to light either toward the source of light (positive phototropism) or away from it (negative phototropism).

Phosphorylation: in biochemistry, phosphorylation is the attachment of a phosphate group to a molecule or an ion. This process and its inverse, dephosphorylation, is common in biology. Protein phosphorylation often activates (or deactivates) many enzymes.

Redox Reactions: redox reactions are very important in living organisms. In a redox reaction, electrons are transferred between species, be they atoms, molecules, ions, free radicals etc. In a reduction reaction, a species gains an electron(s) whereas in an oxidation reaction an electron(s) is lost. In a redox reaction, one species gains an electron (an acceptor) at the same time as another species loses an electron (a donor)—hence the term. An oxidizing agent accepts an electron and is reduced in the process while a reducing agent gives an electron and is oxidized.

RNA: ribonucleic acid is present in all living cells and in many viruses. RNA consists of a long, usually single-stranded chain of alternating phosphate and ribose units, with one of the bases, adenine, guanine, cytosine or uracil bonded to each ribose molecule. Unlike DNA, RNA is more often found in nature as a single-strand folded onto itself, rather than a paired double strand. Cellular organisms use RNA to convey the genetic information (using the letters G, A, U and C to denote the nitrogen bases guanine, adenine, uracil and cytosine), which directs synthesis of specific proteins. Many viruses encode their genetic information using an RNA genome.

Secondary Metabolites: Secondary metabolites in plants are chemicals that, thus far, have no identified role in growth, photosynthesis, reproduction or any other 'primary' function. These chemicals are extremely diverse, indeed, many thousands are known. They may be classified on the basis of chemical structure into three main groups: the terpenes (composed almost entirely of carbon and hydrogen), phenol related compounds (made from simple sugars, containing benzene rings, hydrogen and oxygen) and molecules containing nitrogen. The fact that many secondary metabolites have a specific negative impact upon other organisms, such as grazing animals, insects or bacteria, has led to the hypothesis that secondary metabolites have evolved in plants directly because of their protective value.

Steric hindrance: steric hindrance occurs when the large size of a particular group within a molecule prevents chemical reactions that are observed in related molecules without that group. Although steric hindrance is sometimes a problem, it can also be exploited to change the pattern of the chemistry of a molecule by inhibiting unwanted side-reactions (known as steric protection) or by leading to a preference for one course of stereochemical reaction over another (as in *diastereoselectivity*).

Sublimation: sublimation is said to occur when a substance passes directly from the solid to the vapor phase without going through the intermediate liquid phase (and vice versa).

Supercritical Fluids: any substance, provided it does not break down chemically and physically, can become a supercritical fluid at a temperature and pressure above its **critical point** where distinct liquid and gas phases do not exist. Supercritical fluids have properties between those of a gas and a liquid. As there is no liquid/gas phase boundary, there is no surface tension in a supercritical fluid. By changing the pressure and temperature of the fluid, its properties can be 'tuned' to be more like a liquid or more a gas. The solubility of a supercritical fluid tends to increase with density of the fluid (at constant temperature). Since density increases with pressure, solubility tends to increase with pressure—a property that can be exploited in industrial processes.

Tincture: a tincture is an old-fashioned medical term for a solution of an organic compound in ethanol.

Van Der Waals forces: Van Der Waals forces are forces of attraction that exist between all atoms and all molecules. They arise from electrostatic attraction between temporary dipoles and induced dipoles caused by the movement of electrons within atoms and molecules. Consequently, Van Der Waals forces are much weaker than any other type of molecular interaction. They are only significant at all in atoms and molecules which have no other form of intermolecular attraction, examples being non-polar molecules and the Noble gases, Group 0, in the Periodic Table of elements.

Winchester Bottle or Boston Round: a large glass storage bottle for liquids, sometimes tinted amber to absorb light, used in the chemical industry and in chemical laboratories.

Organic Chemistry: Miracles from Plants

Acknowledgment of Sources for Illustrations

Illustrations for figures are either in the public domain or have been transferred from *'Botanical Miracles, Chemistry of Plants that Changed the World'* by Raymond Cooper and Jeffrey J. Deakin, a CRC publication, in which case they are indicated by an asterisk.

Other sources of illustrations are acknowledged separately in the figure captions throughout the book.

Part I Introduction

Part II Medical Marvels

Part V Euphorics

Part VI Exotic Potions, Lotions and Oil

Part VII Colorful Chemistry; a Natural Palette of Plant Dyes and Pigments

Part VIII Plant Materials

Part IX Plants and the Natural Environment

Acknowledgments

Organic Chemistry in Context: Miracles from Plants would not have been possible without the brilliance of many organic chemists who have dedicated their research over many decades to the challenge of elucidating the chemical structures and properties of the exquisite and complex molecules found in the world of plants. They have left a lasting legacy.

Thanks are also offered for the guidance and patient help of Hilary LaFoe and Sukirti Singh at Taylor and Francis/CRC Press and for inspiration from Raymond Cooper—a fellow chemist and a life-long friend.

Index

Note: Page numbers in *italics* indicate a figure and page numbers in **bold** indicate a table on the corresponding page.

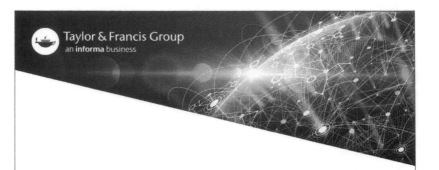